Hilbert's third problem: scissors congruence

C-H Sah

State University of New York at Stony Brook

Hilbert's third problem: scissors congruence

Pitman Advanced Publishing Program

SAN FRANCISCO · LONDON · MELBOURNE

FEARON PITMAN PUBLISHERS INC.
6 Davis Drive, Belmont, California 94002

PITMAN PUBLISHING LIMITED
39 Parker Street, London WC2B 5PB
North American Editorial Office
1020 Plain Street, Marshfield, Massachusetts 02050

Associated Companies
Copp Clark Pitman, Toronto
Pitman Publishing New Zealand Ltd, Wellington
Pitman Publishing Pty Ltd, Melbourne

AMS Subject Classifications: (main) 50-02, 50A30, 50B25, 50L05
(subsidiary) 18H10, 20J10, 20H15

Manufactured in Great Britain

ISBN 0 273 08426 7

TO GERHARD HOCHSCHILD

His timely invitation for me to visit Berkeley during the fall quarter of 1976 gave me the opportunity to venture into unfamiliar grounds.

His constant friendship and encouragements enabled me to find my way through an endless amount of cutting and pasting of polyhedra.

Ultimately, his mathematical works in the areas of homological algebra, number theory, Lie algebras, Lie groups, algebraic groups and Hopf algebras furnished me the techniques to tackle the various aspects of the problem.

Preface

In his 1977 Chicago notes on Hilbert's Problems [45], Kaplansky noted two distinguishing features of the third problem. I will amplify these a little: First, the problem could be understood by anyone with a rudimentary acquaintance with geometry. Second, the problem sought a counterexample rather than a general program. Third, the problem was the first to be solved; in fact, it was not among the ten problems formally presented by Hilbert in his address and the solution by Max Dehn a few months later actually preceded the printed appearance of the entire collection of problems. As a result, outside of a few experts, the third problem has been more or less forgotten by the mathematical public. All of these probably contributed to one more feature: the oral exposition on the third problem in the 1974 DeKalb Symposium was not written up in the proceedings [14]. To a large extent, this last feature vanished with the recent appearance of Boltianskii's book [11]. Against such a backdrop, another monograph on Hilbert's third problem would seem to be out of place. However, with a slight change in viewpoint, I believe that there is still much to be learned about the mathematics associated with the third. Aside from the inherent geometry, I see the appearance of many interesting questions in algebra, analysis, combinatorics, homology theory, number theory as well as many others. In my opinion, the third problem is of equal stature with the rest of the problems of Hilbert. The purpose of the present work is to give an account of some of the recent developments associated with the third that are not covered by Boltianskii's book. A bird's eye view of the work is delegated to Chapter 1. In areas beyond my competence, the sketches are quite rough and inaccuracies are no doubt abound. I have included a number of open problems and loose comments to convey the idea that the subject is still quite lively and the present work is definitely not the final words.

I now come to the pleasant task of thanking those who had a hand in the present project. Since a complete list would be too long, I must apologize in advance to those who may have been offended by my sin of omission.

To Dennis Sullivan, his five minute tutorial on Dehn invariants over a cup of coffee during the summer of 1975 introduced me to a beautiful problem; I am too ashamed to admit that other interests and duties kept me from following up the lead until the summer of 1976.

To my colleagues (too numerous to name individually) in Berkeley, Stony Brook and elsewhere, in particular, to the referee of the monstrous [60], their willing ears and eyes as well as helpful and critical comments were valuable guides.

To H. Hadwiger and J. P. Sydler, the ideas in their works definitely shaped my thoughts; my regret is that I have only met them through their published works.

To Børge Jessen and Anders Thorup, for allowing me the free use of their ideas contained in private letters and preprints even though we have never met. They deserve much of the credits for the completion of the present work. Needless to say, I take full responsibility for the mistakes I have introduced in my interpretations of their thoughts.

To Jeff Cheeger, Mikhail Gromov, John Milnor and James Simons, for the interesting geometric problems they raised, and for the geometric tutorials conducted for my benefits.

To Ronald Douglas and other members of Pitman Publishing Limited, for granting me the priviledge to have the present work published in the research notes series.

To the National Science Foundation, the University of California at Berkeley and the State University of New York at Stony Brook for their financial support.

To Analee, Adam, Jason and Deborah for putting up with my many changing moods.

Contents

1 A rapid transit

1. HILBERT'S THIRD PROBLEM.

In Euclidean plane geometry, areas of polygons can be computed through a finite process of cutting and pasting (scissors congruence). A careful analysis was carried out by Hilbert in his monograph on the foundations of geometry [37]. The third problem of Hilbert was posed because Hilbert did not believe a theory of volume can be based on the idea of cutting and pasting. Instead of recounting the interesting historical events, we simply record the statement of the third problem [14; p. 10]:

> In two letters to Gerling, Gauss (Werke, v. 8. p. 241, 244) expresses his regret that certain theorems of solid geometry depend upon the method of exhaustion, i. e., in modern phraseology, upon the axiom of continuity (or upon the axiom of Archimedes). Gauss mentions in particular the theorem of Euclid, that triangular pyramids of equal altitudes are to each other as their bases. Now the analogous problem in the plane has been solved. Gerling also succeeded in proving the equality of volume of symmetrical polyhedra by dividing them into congruent parts. Nevertheless, it seems to me probable that a general proof of this kind for the theorem of Euclid just mentioned is impossible, and it should be our task to give rigorous proof of its impossibility. This would be obtained, as soon as we succeeded in specifying two tetrahedra of equal bases and altitudes which can in no way be split up into congruent tetrahedra, and which cannot be combined with congruent tetrahedra to form two polyhedra which themselves could be split up into congruent tetrahedra.

In contrast with the other problems of Hilbert, the third only sought a counterexample. This was accomplished by Dehn [25] a few months after the problem was posed. Dehn's work extended and corrected earlier works of Bricard [13], see [11; p. 118]. The following two tetrahedra (as well as uncountably many others) can be shown to yield the desired counterexamples:

A = convex closure of (0, 0, 0), (1, 0, 0), (0, 1, 0), (0, 0, 1);

B = convex closure of (0, 0, 0), (1, 0, 0), (0, 1, 0), (0, 1, 1).

They have common volume 1/6 but distinct (first, or codimensional 2) Dehn invariant. In general, a Dehn invariant is any function f of simplices with values in a suitable abelian group so the following conditions hold:

(C) f(S) = f(T) if S and T are congruent simplices; and

(A) f(R) = f(S) + f(T) if S and T are simplices resulting from a simple subdivision of the simplex R.

Volume is such an example. In present day terminology, the (first, or codimensional 2) Dehn invariant of an n-simplex A in $\mathbb{R}^n$, $n \geq 3$, has its value in the abelian group $\mathbb{R} \otimes_{\mathbb{Z}} (\mathbb{R}/\mathbb{Z})$. It is given by the formula:

$$E\Phi^{(2)}(A) = \Sigma \operatorname{vol}_{n-2}(A^{(2)}) \otimes \overline{\theta}_A(A^{(2)})$$

where the summation extends over the n(n+1)/2 codimensional 2 faces $A^{(2)}$ of A; $\operatorname{vol}_{n-2}(A^{(2)})$ is the absolute (n-2)-dimensional volume of $A^{(2)}$ and $\overline{\theta}_A(A^{(2)})$ lies in $\mathbb{R}/\mathbb{Z}$ so that $\theta_A(A^{(2)})\pi/2$ is the radian measure of the interior dihedral angle at the face $A^{(2)}$ of A. This formulation appeared in the book of Hadwiger [31]. In a letter as well as [28], Debrunner pointed out that Nicolletti [56] had already invented this tensor product formulation in 1915, compare MacLane [50; p. 172]. The main feature in the tensor product is the fact that $\theta_A(A^{(2)})$ can be replaced by 0 whenever it is rational. It is then easy to see that $E\Phi^{(2)}(B) = 0$. A little more work shows that $E\Phi^{(2)}(A) \neq 0$. Property (C) is clear for $E\Phi^{(2)}$. A little bit of work is required to verify property (A). Granting these, A and B then furnish the desired counterexample.

2. ABSTRACT SCISSORS CONGRUENCE.

The abstract scissors congruence data consist of a distinguished family of nonempty subsets (to be called the n-simplices where n is to be interpreted as the dimension) of a nonempty set X and a specified equivalence relation (called congruence) among the n-simplices. We need a few definitions before stating the addition condition to be satisfied by the n-simplices.

Two n-simplices A and B are said to be interior disjoint if the following conditions hold:

(D1) $A \cap B$ contains no n-simplices, and

(D2) if C is an n-simplex contained in $A \cup B$, then $C \subset A$ if and only if $C \cap B$ contains no n-simplices.

A polyhedron P is understood to be a finite pairwise interior disjoint union of n-simplices. The concept of interior disjoint union can then be extended to polyhedra with n-simplices replaced by nonempty polyhedra. We will use $\coprod$ to denote pairwise interior disjoint unions (as well as direct sum when there is no chance of confusion). We omit the proof of the following elementary result:

Lemma 2.1. If P, Q and R are polyhedra with $P \coprod R = Q \coprod R$, then $P = Q$.

If A, B, C are n-simplices with $A = B \coprod C$, we say A is simply subdivided into B and C. If $P = \coprod P_i$ is a polyhedron where each P_i is an n-simplex, then a subdivision of P is understood to be a finite succession of simple subdivisions such that each simple subdivision is performed on one of the n-simplices exhibited in the preceeding step. For example, the first step may be a simple subdivision of P_1 into Q and R; the second step may be a simple subdivision of Q or R or any P_j with $j > 1$; and so on.

To complete the abstract scissors congruence data, we impose the following condition:

(S) Let A and B be n-simplices. Then there is at least one subdivision of A, say $A = \coprod_{1 \leq i \leq t} A_i$, such that $A \cap B = \coprod_{j \in J} A_j$ for some $J \subset \{1, \ldots, t\}$.

We omit the proof of the following elementary result:

Lemma 2.2. Let $P = \coprod_i A_i = \coprod_j B_j$ be representations of the polyhedron P with A_i, B_j denoting n-simplices. Then there is a common subdivision $P = \coprod_s C_s$ so that each A_i and each B_j is the union of suitable C_s's.

Two polyhedra P and Q are said to be scissors congruent (denoted by $P \sim Q$) if $P = \coprod_i P_i$, $Q = \coprod_i Q_i$ with P_i congruent to Q_i, $1 \leq i \leq t$. They are said to be stably scissors congruent (denoted by $P \overset{s}{\sim} Q$) if there exists a finite collection of polyhedra R_i, S_i, T_i so that:

$$T_0 = P;\ T_t = Q;\ (T_i \amalg R_i) \sim (T_{i-1} \amalg S_i) \text{ and } R_i \sim S_i \text{ for each } i.$$

Axiom (S) ensures that our congruences are equivalence relations. The abstract problems are:

> Find reasonable necessary and sufficient conditions for two polyhedra P and Q to be scissors congruent or to be stably scissors congruent.

In most of the cases considered by us, the congruence relation arises from a group G of motions of the underlying space. We will then speak of G-scissors congruence or stable G-scissors congruence and write $\sim_G$ and $\overset{s}{\sim}_G$. When the meaning is clear, G will be omitted. In the literature scissors congruence is often called equidecomposability while stable scissors congruence is often called equicomplementability. We now consider examples.

First Example. Let G be the group of all permutations of an infinite set X. Let $n \geq 0$ be an integer. Define n-simplices as arbitrary (n+1)-element subset of X. When n = 0, interior disjointness coincides with set theoretic disjointness. When n > 0, two n-simplices are never interior disjoint. (S) is clear and cardinality is a complete invariant. This example shows that interior disjointness and set theoretic disjointness may be independent.

Second Example. Let X be the affine n-space over the field $\mathbb{R}$ of all real numbers. Let G denote the group generated by all translations and all homotheties (uniform expansions and contractions with respect to an origin). Define an n-simplex to be any n-dimensional parallelopiped with its edges parallel to a fixed set of coordinate axes. It is easy to see that we have scissors congruence data. By using the fact that the greatest common divisor of all $m^n - 1$, $m \in \mathbb{Z}^+$, is 1, it is easy to see that every n-simplex is stably G-scissors congruent to the empty set.

Third Example. Here we follow Hilbert [37]. Let $X = k^2$ be the affine plane based on any ordered field k. Equip X with the usual positive definite inner product and let G be the group of all affine automorphisms of X fixing this inner product. Define 2-simplices to be triangles. It is now easy to check that we have scissors congruence data. The following diagrams show that two parallelograms with the same height and same base are stably G-scissors congruent. In fact, G can be replaced by the subgroup of all translations.

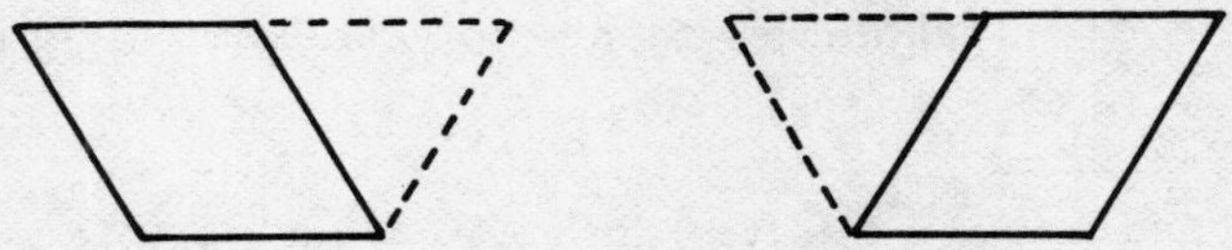

Theorem 2.3. With the preceding notations, two polygons P and Q in X are stably G-scissors congruent if and only if they have the same area. If k is Archimedean, then stable G-scissors congruence can be replaced by G-scissors congruence.

A proof can be based on a combination of the preceding figures together with the following figures. We leave the details to the reader who prefers rigor over oriental puzzles.

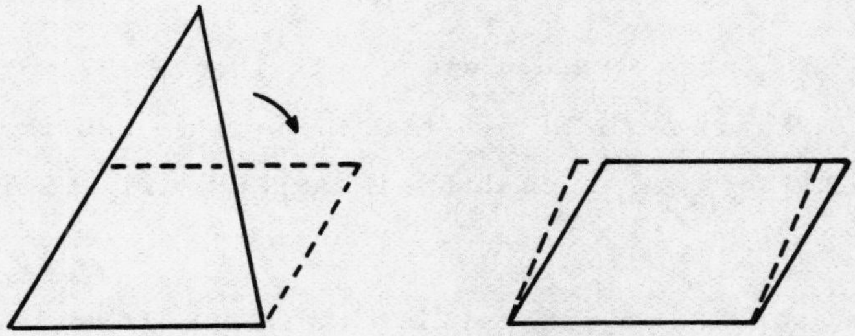

Part of Theorem 2.3 goes back to Euclid. When $k = \mathbb{R}$, Theorem 2.3 has been attributed to F. Bolyai (1832) and P. Gerwien (1833), [11; p. 50]. In a private communication, B. Jessen indicated that a proof was found by William Wallace in 1807 [71]. The group G in Theorem 2.3 can be replaced by the subgroup G_0 generated by all translations together with -Id with respect to an origin. Boltianskii [11; p. 87] showed that G_0 is the minimal group for the validity of Theorem 2.3. The following is due to Hilbert [37].

<u>Theorem 2.4.</u> If k is nonarchimedean, then there are polygons in $X = k^2$ which are stably G-scissors congruent, but not G-scissors congruent.

Theorem 2.4 extends to higher dimensions. For a proof, see Theorem 2.3.5. We note that the inner product is used in proving Theorem 2.4. The algebraic questions related to inner products and ordered fields led to Hilbert's 11-th and 17-th problems, thence to geometric algebra.

3. <u>THEOREM OF ZYLEV</u>.

In some sense, Theorem 2.4 typifies the cases where stable scissors congruence does not imply scissors congruence. We say a polyhedron P is <u>larger than twice the polyhedron</u> Q when the following condition holds:

> Let $P_1 \subset P$ and $Q_1 \subset Q$ be arbitrary subpolyhedra with $P_1 \sim Q_1$. Then there is a subpolyhedron $P_2 \subset P$ such that P_2 is interior disjoint from P_1 and such that $P_2 \sim Q$. ($P_1 = \emptyset = Q_1$ is allowed.)

The following is an abstract version of a theorem of Zylev [76].

<u>Theorem 3.1.</u> Assume scissors congruence data have been given and the following additional conditions hold:

(VP) If $P \subset Q$ are polyhedra with $P \sim Q$, then $P = Q$.

(VA) If A, B are n-simplices, then there is a subdivision $A = \coprod_{1 \leq i \leq t} A_i$ such that B is larger than twice A_i, $1 \leq i \leq t$.

Then stable scissors congruence implies (therefore, is equivalent with) scissors congruence.

Proof. Conditions (VP) and (VA) are the abstract substitutes for a volume invariant under the given congruence. (VP) is clearly necessary for the conclusion; it is an abstraction of the condition that n-simplices should have positive volume. (VA) is not necessary for the conclusion (for example, an affine line over a nonarchimedean ordered field does not satisfy (VA) but the conclusion of Theorem 3.1 does hold); it is an abstract form of the archimedean property of volume.

In order to prove our assertion, it is enough to show:

$$\text{Let } P, Q, R \text{ and } S \text{ be polyhedra with } (P \amalg R) \sim (Q \amalg S) \text{ and } R \sim S. \text{ Then } P \sim Q. \tag{3.2}$$

Applying a scissors congruence, we may assume $P \amalg R = Q \amalg S$ in (3.2). Using (VP), we may assume $P \neq \emptyset \neq Q$. Similarly, we may assume $R \neq \emptyset \neq S$. Using (VA) and (S), we may assume $R = \amalg R_i$, $S = \amalg S_i$, R_i congruent to S_i and such that P is larger than twice of R_i (hence S_i), $1 \leq i \leq t$. The proof will now proceed by induction on t. We will carry out scissors congruences of $P \amalg R$ with itself. It should be noted that a scissors congruence is not necessarily a set theoretic bijection.

Applying (VA), we can find a polyhedron T_1 contained in P, interior disjoint from $S_1 \cap P$ and $T_1 \sim S_1$. Fix a scissors congruence between T_1 and S_1 and "exchange" $S_1 \cap R$ with its image in T_1 under the chosen scissors congruence (the exchange amounts to redefining congruences). By leaving the complementary polyhedron (of the exchanged pieces with respect to $P \amalg R$) alone, we have a scissors congruence of $P \amalg R$ with itself. We therefore have $P \amalg R = P' \amalg R'$ where $P \sim P'$, $R' \sim R$ and $R' = \amalg R'_i$ with $R'_i \sim R_i \sim S_i$ and $S_1 \subset P'$. We next perform a scissors congruence on $P' \amalg R'$ exchanging S_1 and R'_1 and leaving the complement polyhedron (of the exchanged pieces with respect to $P' \amalg R'$) alone. This gives $P' \amalg R' = P'' \amalg R''$ with $P' \sim P''$, $R' \sim R''$, $R'' = \amalg R''_i$, $R''_i \sim R'_i$ and $R''_1 = S_1$. By Lemma 2.1,

$$P'' \amalg (\amalg_{i>1} R''_i) = Q \amalg (\amalg_{i>1} S_i)$$

An examination of the argument shows that we did not use the congruence between R_i and S_i. Everything depended only on $R_i \sim S_i$. Since (VA) is invariant under scissors congruence and since $P'' \sim P$, $R_i'' \sim R_i \sim S_i$ and P is still larger than twice of S_i for each i, induction applies. We therefore have $Q \sim P'' \sim P$. Q. E. D.

The following diagrams summarize the steps leading to induction:

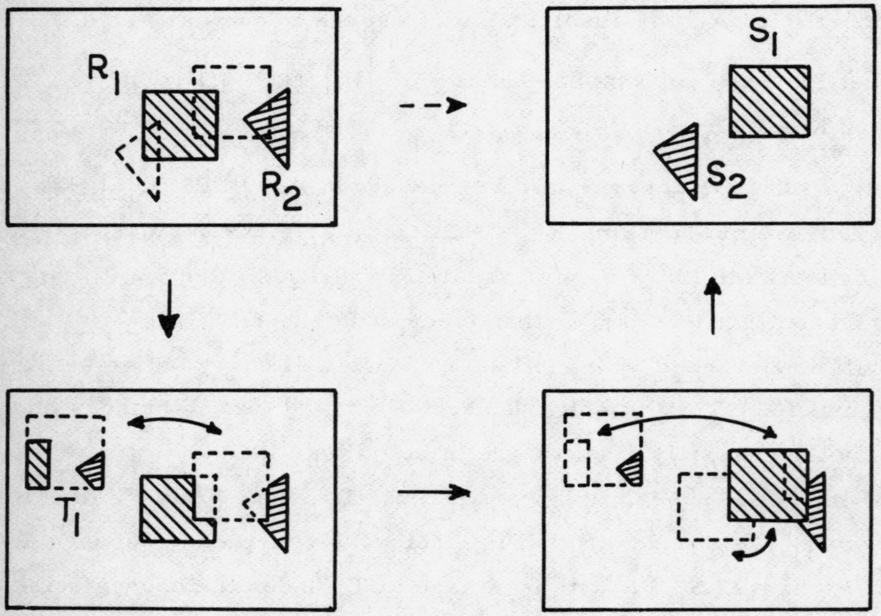

In the top two diagrams, the unshaded polyhedra represent P and Q resptively. In the bottom two diagrams, the unshaded polyhedra represent P' and P'' respectively. These diagrams illustrate the case of G-scissors congruence. However, with some changes in notations, they can be made to work for general scissors congruence as well. As indicated in the proof, the modification amounts to giving names to the various scissors congruences, the domains and the targets. For our purposes, G-scissors congruences will occupy our attention after this chapter.

Stable scissors congruence has the advantage that it is more easily described in terms of abelian groups. For the generators, we take the symbols [A], one for each n-simplex A. We then impose the defining relations:

(RC) [A] = [A'] if A and A' are congruent n-simplices;

(RS) [A] = [B] + [C] if A is simply subdivided into B and C.

The abelian group so obtained is denoted by $\mathcal{P}^n(X)$ or $\mathcal{P}(X)$. When dealing with G-scissors congruence, we also write $\mathcal{P}^n(X, G)$ or $\mathcal{P}(X, G)$. If $P = \coprod P_i$ is a polyhedron with P_i denoting n-simplices, $1 \leq i \leq t$, then [P] denotes $\Sigma_i [P_i]$.

In order to give an interpretation of the elements of $\mathcal{P}(X)$, we let X_i be the disjoint union of i copies of X with $i = 1, 2, \ldots, \infty$. The scissors congruence data on X can be extended in an obvious way to X_i. It is then clear that $\mathcal{P}(X_i)$ and $\mathcal{P}(X)$ are naturally isomorphic. Each element of $\mathcal{P}(X_\infty)$ can be written as [P] - [Q] with suitable polyhedra P, Q (empty polyhedron is allowed) in X_∞. In general, the scissors congruence data is said to be <u>infinite</u> when the following condition holds:

(INF) Let P be any polyhedron and let A be any n-simplex. Then there is a polyhedron B with B interior disjoint from P and with $B \sim A$.

When the given scissors congruence data is infinite, each element of $\mathcal{P}(X)$ can be written as [P] - [Q] with P, Q denoting suitable polyhedra in X. Moreover, $P \overset{s}{\sim} Q$ holds if and only if [P] = [Q]. For this reason, [P] will be called the stable scissors congruence class of P even when our data is not infinite. We note that the data on X is infinite if and only if the data on X_i is infinite for any finite i.

<u>Proposition 3.3</u>. Condition (EQ) below holds for the scissors congruence data on X if and only if it holds for the extended data on X_i, $1 \leq i \leq \infty$.

(EQ) stable scissors congruence is equivalent with scissors congruence.

Proof. Sufficiency holds because X can be identified with one of the i copies of X in X_i. Since (EQ) is a statement of finite character, the necessity for X_∞ is a consequence of the necessity for X_i, $i \in \mathbb{Z}^+$. By iteration, it is enough to consider the case of $X_2 = X \times \{0\} \coprod X \times \{1\}$. We must show:

$$\text{If } A = B \coprod C,\ A' = B' \coprod C' \text{ are polyhedra in } X_2 \text{ with } A \sim A' \text{ and } C \sim C', \text{ then } B \sim B'. \tag{3.4}$$

Of course, we assume (3.4) holds for X in place of X_2. We now show that there is no loss in generality if we add the assumption $C = C'$ in (3.4). To see this, we apply a suitable scissors congruence in X_2 and assume $A = A'$. We therefore have:

$$B = (B \cap B') \coprod (B \cap C'),\ B' = (B' \cap B) \coprod (B' \cap C);$$
$$C = (B' \cap C) \coprod (C \cap C'),\ C' = (B \cap C') \coprod (C' \cap C).$$

In view of the forms of B and B', it is enough to show $B \cap C' \sim B' \cap C$. In view of the assumption on C, C' and their forms, it is clear that we can add the assumption $C = C'$ in the verification of (3.4). Writing C and C' out as $R \times \{0\} \coprod S \times \{1\}$ and $R' \times \{0\} \coprod S' \times \{1\}$, (3.4) becomes:

$$\text{If } ((P \coprod R) \times \{0\}) \coprod ((Q \coprod S) \times \{1\}) \sim ((P' \coprod R') \times \{0\}) \coprod ((Q' \coprod S') \times \{1\}),$$
$$\text{with } R \sim R',\ S \sim S', \text{ then } (P \times \{0\}) \coprod (Q \times \{1\}) \sim (P' \times \{0\}) \coprod (Q' \times \{1\}).$$

We now fix a scissors congruence between A and A' in the preceding normalization. If a subpolyhedron $R_1 \times \{0\}$ of $R \times \{0\}$ is carried into $R' \times \{0\}$, then we can use (EQ) for X and cancel R_1 and its image in R'. The same argument applies to S and S'. We have reached the following stage:

> There is a scissors congruence between $((P \coprod R) \times \{0\}) \coprod ((Q \coprod S) \times \{1\})$ and $((P' \coprod R') \times \{0\}) \coprod ((Q' \coprod S') \times \{1\})$ such that the image of $R \times \{0\}$ (resp. $S \times \{0\}$) is interior disjoint from $R' \times \{0\}$ (resp. $S' \times \{0\}$). If $R \sim R'$, $S \sim S'$, then $(P \times \{0\}) \coprod (Q \times \{1\}) \sim (P' \times \{0\}) \coprod (Q' \times \{1\})$ holds when (EQ) holds for X. (3.5)

Let σ denote the scissors congruence mentioned in (3.5). Let ρ denote a scissors congruence between R and R'. Write $R = R_1 \coprod R_2$ so that:

$$\sigma(R_1 \times \{0\}) \subset S' \times \{1\} \text{ and } \sigma(R_2 \times \{0\}) \subset (P' \times \{0\}) \coprod (Q' \times \{1\})$$

We can now modify σ by exchanging $\rho(R_1) \times \{0\}$ and $\sigma(R_1 \times \{0\})$. This new scissors congruence σ' no longer has the property mentioned in (3.5) for σ. But $\sigma'(R \times \{0\})$ is interior disjoint from $S' \times \{1\}$. The argument leading to (3.5) can be repeated for σ' since it involves cancellations of corresponding subpolyhedra of $R \times \{0\}$ and $R' \times \{0\}$ (resp. of $S \times \{1\}$ and $S' \times \{1\}$). This means that (3.5) can be strengthened so that $\sigma(R \times \{0\})$ is interior disjoint from $(R' \times \{0\}) \coprod (S' \times \{1\})$. With this improvement at hand, we can repeat with $S \times \{1\}$. The net effect is that the map σ in (3.5) can be assumed to satisfy:

$\sigma(C)$ is interior disjoint from C'.

In reaching this, the original C and C' have been diminished as well as changed under scissors congruences applied to A and A'. Since the diminishing process is done under scissors congruences on A and A', it amounts to taking parts of A and A' and leave them alone; therefore, we have not lost any of our assumptions. We have reached the stage:

$A = B_1 \coprod B_2 \coprod C$, $A' = B_1' \coprod B_2' \coprod C'$, and σ is a scissors congruence between A and A' so that $\sigma(B_2) = C'$ and $\sigma(C) = B_2'$.

It follows that $\sigma(B_1) = B_1'$ holds above. Since $C \sim C'$ by assumption, we have $B = B_1 \coprod B_2 \sim B_1' \coprod B_2' = B'$ Q.E.D.

<u>Corollary 3.6.</u> Conditions (VP) and (VA) of Zylev's theorem are valid for scissors congruence data on X if and only if they are valid for any X_i.

<u>Proof</u>. As in Proposition 3.3, sufficiency is clear. Assume (VP) and (VA) hold for X. Since (VP) involves only n-simplices, its validity on X extends to X_i. As for (VA), we use Zylev's theorem together with Proposition 3.3.

<u>Remark 3.7.</u> It would be interesting to know if the validity of (VA) for X_i is actually a direct consequence of its validity on X. The preceding proof detours around the direct proof.

Remark 3.8. If X were a manifold, then X_i can be viewed as an i-sheeted covering space of X (in the sense of scissors congruence). Aside from the archimedean property, (VA) also signifies the homogeneity property of X. As an illustration, let X be the unit circle. Take 1-simplices to mean arcs of the circle X. Define congruence by using the infinite cyclic group generated by an irrational rotation. Then (VP) and (VA) are both valid and $\mathcal{P}(X)$ is quite large. In this example, G is not transitive on X, but is ergodic on X. The properties (VP) and (VA) are independent. To see this, we take X to be $\mathbb{R}^n$ with the usual definition of n-simplices. Define congruence by taking G to be $\{1\}$ or to be the group of all uniform expansions and contractions with respect to the origin. These two examples imply independence.

4. CONCRETE SCISSORS CONGRUENCES.

We now survey the scissors congruence problems where the underlying set X is among the classical geometries: affine space, spherical space, and hyperbolic space. Among these, the most complete results concern the affine space. For this case, there are choices for the group G of motions. In all cases, we are dealing with G-scissors congruences.

Let $\mathfrak{m}$ be an n-dimensional vector space over an ordered field k. Let G be any subgroup of the group $A_k(\mathfrak{m})$ of all affine transformations of $\mathfrak{m}$. The group $A_k(\mathfrak{m})$ is the semidirect product of the normal subgroup $T_k(\mathfrak{m})$ of all translations by the subgroup $GL_k(\mathfrak{m})$ of all k-linear automorphisms of $\mathfrak{m}$. An n-simplex is defined to be the convex closure of any n+1 affinely independent points of $\mathfrak{m}$. The ordering in k is needed to define convexity as well as simple subdivision. Using the affine structure, we can check that we have scissors congruence data with congruence defined by any subgroup G of $A_k(\mathfrak{m})$. In order to obtain (VP), the simplest way is to demand that G preserves volume. For affine spaces, volume can be based on determinant. To obtain G-invariance, it is sufficient (as well as necessary) that G be a subgroup of $SA_k^{\pm}(\mathfrak{m}) = T_k(\mathfrak{m})SL_k^{\pm}(\mathfrak{m})$ where $SL_k^{\pm}(\mathfrak{m})$ is the group of all k-linear automorphisms of $\mathfrak{m}$ with determinant ± 1. In order to obtain (VA), G must be fairly large and k must not be totally arbitrary. We assume $G \supset T_k(\mathfrak{m})$.

Affine Spaces. We consider the cases according to the choice of G.

$G = SA_k^{\pm}(\mathfrak{m})$. From determinant theory, two n-simplices are G-congruent if and only if they have the same absolute volume. The absolute volume is a complete and independent invariant for G-scissors congruence as well as stable G-scissors congruence. This case is therefore elementary.

$G = T_k(\mathfrak{m})$. This is called the translational case. In view of the preceding case, we often call the present case the affine case. It could also be called the Minkowski case (suggested by Chern) or the Hadwiger case (in honor of its originator). The general translational case was first formulated in early 1950's when Hadwiger recognized that certain aspects of the third problem depended only on having all the translations. For example, Zylev's theorem was proved after Hadwiger had proved it in the translational case with $k = \mathbb{R}$ and n arbitrary (thereby extending the 3-dimensional result of Sydler). Under the assumption that $k = \mathbb{R}$, the translational cases for n = 2 and 3 were completely settled by Hadwiger-Glur [34] in 1951 and by Hadwiger [33] in 1968 respectively. The general problem of stable $T_k(\mathfrak{m})$-scissors congruence was completely settled in 1972-73 by B. Jessen and A. Thorup, the final manuscript was submitted for publication in December of 1977 [44]. In our ignorance, this result was rediscovered in 1976-77 by a slightly different path in [60]. In combination with Hadwiger-Zylev theorem, we have a complete system of invariants (called the Hadwiger invariants) for translational scissors congruence over any Archimedean field k in any dimension n. Chapters 2 through 5 are devoted to the translational case. The principal point is the existence of a complicated k-vector space structure on the groups $\mathcal{P}(k^n, T(n;k))$. This vector space structure is intimately tied up with the similarity automorphisms of k^n. The similarity action on $\mathcal{P}(k^n, T(n;k))$ is highly nonlinear. The Hadwiger invariants are not independent. The study of the relations among them is called the syzygy problem; it is studied in Chapter 5 and solved only for the first syzygy. The higher syzygy problem is open for $n > 3$ and appears to be connected with the Eilenberg-MacLane cohomology of orthogonal groups with certain wellknown nontrivial local coefficients (this seems to be the first time such types of cohomology have appeared--outside of the trivial coefficients and the arithmetic situations). As it will become apparent, the translational case is essentially a problem in geometric algebra.

For the other cases, k will be assumed to be Archimedean ordered and square root closed field (i.e., $k^+ = (k^+)^2$). Equivalently, k is a square root closed subfield of $\mathbb{R}$. In the text, many of the results have been stated and proved without the Archimedean assumption.

Spherical Spaces. The space is the n-sphere $S(k^{n+1})$ consisting of all unit vectors in the Euclidean space k^{n+1} (with the usual positive definite k-valued inner product). The group G of motions is $O(n+1;k)$ consisting of all orthogonal transformations of k^{n+1}. An n-simplex is defined to be the (geodesic) convex closure of n+1 k-linear independent unit vectors. These then define spherical scissors congruence data; in fact, it makes sense without either the Archimedean or the square root closure assumption. However, using the square root closure assumption, we can make a ring out of all the groups $\mathcal{P}S^{n+1} = \mathcal{P}(S(k^{n+1}), O(n+1;k))$, $n \geq -1$. By using the central symmetry -Id, $\mathcal{P}S^{2t+1}$ reduces to $\mathcal{P}S^{2t}$ modulo torsion. This signifies the combinatorial aspect of Gauss-Bonnet volume calculation. Basic geometric facts and the spherical version of Dehn invariants can be combined to yield a Hopf algebra structure on a certain quotient ring of our ring. This can be combined with the Hopf-Leray theorem to yield a multiplicative cancellation theorem modulo torsion for the spherical scissors congruence problem for Archimedean k. Such a result is unknown in either the Euclidean or the hyperbolic case. When $n \leq 2$, volume is a complete invariant for the scissors congruence problem. When $n > 2$, the Dehn invariants are necessary, but the sufficiency is open. The material in Chapter 6 on the spherical case has its origin in some ideas communicated to us in private letters from B. Jessen and A. Thorup plus a few vague feelings in [60]. Historically, the spherical case rests with the spherical excess formula for the area of a spherical triangle--long after Euclid's area calculation.

Euclidean Spaces. This is actually a special case of the affine spaces. We identify the affine n-space $\mathfrak{M}$ with the space k^n of all column vectors of length n (equipped with the usual positive definite k-valued inner product). The group G of motions is E(n;k) consisting of all rigid motions. It is the semidirect product of T(n;k) and O(n;k). For $n \leq 2$, volume is a complete and independent invariant for the scissors congruence problem. For $n > 2$, the Dehn invariants are necessary conditions (for n = 3, this solves the third problem of Hilbert). When n = 3, the beautiful theorem of Sydler [67] shows the completeness of the Dehn invariants (including volume). Shortly after, the proof due to Sydler was simplified by B. Jessen [40]; indeed, Jessen also determined the range of the Dehn invariants for n = 3. Moreover, using the results from the translational case, Jessen extended the result to n = 4. The sufficiency of the Dehn invariants for $n > 4$ is still pending. As in the spherical case, we can combine $\mathcal{P}E^n = \mathcal{P}(k^n, E(n;k))$, $n \geq 0$, to form a ring. The various forms of the Dehn invariants then convert the ring into (right) comodule for the Hopf algebra constructed in the spherical case. It seems likely that the ring so constructed is actually an integral domain (or even a Hopf algebra), but this is in the conjectural stage. In any event, a theoretic answer can be formulated (trivially) in terms of O(n;k)-invariants on the k-vector space dual to $\mathcal{P}(k^n, T(n;k))$. The difficulty lies with the identification of the Dehn invariants in this setting. In a vague sense, this is related to the syzygy problem. The material in chapter 7 is essentially a rehash of known results. We have restructured the foundation (due to Hadwiger [31]) by incorporating the materials from earlier chapters.

Hyperbolic Spaces. Let $k^{1,n}$ be the space of all column vectors of length n+1; it is equipped with the inner product of signature (1, n) corresponding to the quadratic form $-X_0^2 + \Sigma_{1 \leq i \leq n} X_i^2$. The hyperbolic n-space $\mathfrak{H}^n = \mathfrak{H}^n(k)$ can be modelled as the set of all vectors v in $k^{1,n}$ with $\langle v, e_0 \rangle > 0$ and with $\langle v, v \rangle_{1,n} = -1$. The group G of motions is then the subgroup $\Omega(1, n;k)$ of index 2 in O(1, n;k) stabilizing $\mathfrak{H}^n$ (it is also the kernel of the spinor norm map). An n-simplex in $\mathfrak{H}^n$ is defined to be the (geodesic) convex closure of n+1 k-linearly independent points of $\mathcal{H}^n$. As in the spherical case, these define the hyperbolic scissors congruence data. In contrast to the Euclidean and the spherical cases, -Id is not present in G. Furthermore, the signature restriction prevents us from introducing a product structure. We still have Dehn invariants so that the group formed out of all $\mathcal{P}(\mathcal{H}^n(k), \Omega(1,n;k))$ $= \mathcal{P}\mathcal{H}^n$ is still a (right) comodule for the Hopf algebra constructed in the spherical case. For $n \leq 2$, volume is still a complete scissors congruence invariant. This latter result is attributed to Bolyai in Moise [53]. We have constructed a proof in Chapter 8 by following Gauss in his proof of the defect formula for the area of a hyperbolic triangle. The idea is to attempt to find a combinatorial Gauss-Bonnet theorem in the sense of scissors congruence. Hopefully, this will establish a direct link between the groups $\mathcal{P}\mathcal{H}^{2i}$ and $\mathcal{P}S^{2i}$. The details are still pending. As a result, Chapter 8 is very tentative. We hope to have more to say about this in the near future.

With the exception of the elementary case where X is affine and G is $SA_k^{\pm}(\mathfrak{m})$, the remaining cases are such that X admits a G-invariant metric. Whenever this happens, Dehn type invariants can be defined in terms of volume and angle. The cases where X admits only a G-invariant volume but no G-invariant metric is largely unexplored. A few such examples are mentioned in Chapter 8 in connection with the hyperbolic case. One of the common features in all the cases described so far is the presence of convexity. We conclude the present section with an old result common to the spherical, the Euclidean and the hyperbolic spaces. In each of these, the group G has a unique subgroup G^+ of index 2 consisting of the orientation preserving motions of the underlying space. The result then asserts that G-scissors congruence is the same as G^+-scissors congruence.

Theorem 4.1. Let (X, G) denote the scissors congruence data in spherical, Euclidean, or hyperbolic spaces. Let P and Q be polyhedra in X which are symmetric with respect to a hyperplane. Then P and Q are G^+-scissors congruent. Consequently, (stable) G-scissors congruence is equivalent to (stable) G^+-scissors congruence.

Proof. Writing P as an interior disjoint union of n-simplices, we can take P and Q to be n-simplices which are mirror images of each other. Let the vertices of P be $x_0, \ldots, x_n$ and let P_i be the codimensional 1 face of P opposite x_i. Let z be the center of the inscribed sphere of P so that z is the intersection of the hyperplanes which bisect the interior dihedral angles at the intersection of P_i and P_j, $i \neq j$. Let z_i be the foot of the perpendicular from z to P_i. We then have $P = \coprod_{0 \leq i < j \leq n} A_{i,j}$, where $A_{i,j}$ is the geodesic convex closure of z, z_i, z_j and $P_i \cap P_j$. $A_{i,j}$ is symmetric with respect to the hyperplane determined by z and $P_i \cap P_j$. If $Q = \sigma P$ where σ is a suitable hyperplane symmetry, then $Q = \coprod \sigma A_{i,j}$. Since $A_{i,j} = \rho_{i,j} A_{i,j}$ for a hyperplane symmetry $\rho_{i,j}$ and $\sigma \circ \rho_{i,j} \in G^+$, we have $P \sim Q$. The last assertion follows from this plus the fact that G is generated by hyperplane symmetries. Q. E. D.

Remark 4. 2. In the proof of the preceding theorem, the Archimedean assumption on k is not needed. We need to have orthogonality and the concepts of distances and angles in the plane. We also need to be able to bisect an arbitrary angle. These are available when k is any ordered field which is square root closed. The models we used in the spherical and hyperbolic case allow us to do everything in the setting of geometric algebra so that no difficulty arises. Theorem 4. 1 is often attributed to Gerling. Gerling used the center of the circumscribed sphere (for Euclidean 3-space). According to a private letter from B. Jessen, this construction seems to go back to Legendre. The construction based on the center of the inscribed sphere was used by J. B. Durrande in 1815 [75; p. 948].

Proposition 4. 3. Let (X, G) denote the scissors congruence data in one of the spherical, Euclidean, or hyperbolic n-spaces over an ordered, square root closed field k. The abelian group $\mathcal{P}^n(X, G)$ is then 2-divisible.

Proof. It is enough to show the 2-divisibility of each of the generators of $\mathcal{P}^n(X, G)$. Since a polyhedron symmetric with respect to a hyperplane is clearly 2-divisible in the sense of scissors congruence, the proof of Theorem 4. 1 implies the 2-divisibility of the generators. Q. E. D.

Remark 4.4. In the Euclidean case, $\mathcal{P}^n(X, G)$ is actually a k-vector space by virtue of the results from the translational case. These results use only the assumption that k is an ordered field. For special choices of k, infinite divisibility fails for both the spherical and the hyperbolic cases. In general, torsionfreeness is an open problem. More precisely, if $k = \mathbb{R}$ and $n > 2$, then both divisibility and torsionfreeness are open in the spherical as well as the hyperbolic case. These appear to be connected with number theoretic questions related to the analytic functions appearing in the calculation of volume.

5. MISCELLANEOUS REMARKS.

Aside from the open problems already mentioned, a number of other related problems are posed in the later chapters of the text. The choices are necessarily biased. A number of other interesting topics related to cutting and pasting have been omitted, partly due to lack of time or space and mostly due to our incompetence. The references are by no means complete. We had originally planned to have a complete bibliography. It was soon clear that such a task would be impossible--aside from the sheer size of such a task, there is the problem that references before 1900 are often not accessible. The union of the bibliographies in Hadwiger's book [31], in Kaplansky's notes [45] and in Boltianskii's book [11] cover essentially all the published works connected with Hilbert's third problem dating from 1896 to approximately 1976. Since a large part of the present work is concerned with the mathematics surrounding the third problem, we have included a large number of references which did not concern itself with the third problem but did contain some materials which were useful to us. A number of the references were listed but were not available to us. We learned about them through the kind letters from B. Jessen and H. E. Debrunner. For an evaluation of Dehn's work, the article of Magnus [51] is a must. As a general rule, there is no substitute for original articles, we apologize for having broken this rule many times during the preparation of this monograph because of inaccessibility as well as our ignorance.

2 Affine scissors congruence data

The main task in the present chapter is to lay a foundation for the G-scissors congruence problem when the underlying space X is the affine space of dimension n. The main results are already contained in the book by Hadwiger [31; Chapters 1 and 2]. Some of the results and proofs have been modified in order to bring out the algebraic and the combinatorial aspects of affine geometry. These furnish us the techniques to keep track of the geometry in higher dimensions when visualization becomes blurred.

1. AFFINE POLYHEDRA.

Let $\mathfrak{M}$ be an n-dimensional vector space over an ordered field k. For any subset A of $\mathfrak{M}$, the convex closure of A is denoted by ccl(A). The affine subspace spanned by A is denoted by Asp(A) and the linear subspace parallel to Asp(A) is denoted by Lsp(A). It follows that:

$$\mathrm{Asp}(A) = \{\Sigma_i \alpha_i x_i \mid \Sigma_i \alpha_i = 1,\ x_i \in A,\ \alpha_i \in k \text{ is 0 for almost all } i\};$$
$$\mathrm{Lsp}(A) = \{\Sigma_i \alpha_i x_i \mid \Sigma_i \alpha_i = 0,\ x_i \in A,\ \alpha_i \in k \text{ is 0 for almost all } i\}.$$

When $A = \{x_0, \ldots, x_m\}$ and $\dim_k \mathrm{Lsp}(A) = m$, ccl(A) is called a geometric m-simplex or simply an m-simplex. In such a case, the m+1 vertices, $x_0, \ldots, x_m$ are said to be affinely independent points of $\mathfrak{M}$, we sometimes drop the word affinely when there is no chance of confusion.

Let P and Q be geometric simplices of dimensions p and q respectively. P and Q are said to be interior disjoint if $\dim_k \mathrm{Lsp}(P \cap Q) < \min(p, q)$. In contrast to the usual convention, $\mathrm{Lsp}(\emptyset) = \emptyset$ and has k-dimension -1. Using the fact that a consistent system of linear equations has a solution in the field generated by the coefficients, "dimension" and "interior disjointness" are invariant under ordered field extensions. With the present definition restricted to n-simplices, we have G-scissors congruence data for any group G of affine transformations of $\mathfrak{M}$.

2. NONHOMOGENEOUS FORMULATION.

In order to deal with scissors congruence, it is important to have a precise notation. For the affine case and the Euclidean case, the nonhomogeneous description turns out to be convenient. Indeed, certain manipulations within this frame work are critical for our solution in the affine case.

Let $y_0, \ldots, y_m \in \mathfrak{m}$. Set

$$y_0 + /y_1/\ldots/y_m/ = \text{ccl}\,\{y_0, y_0+y_1, \ldots, y_0+\ldots+y_m\}.$$

When $y_0 = 0$, we write $/y_1/\ldots/y_m/$. In general, $y_0, \ldots, y_m$ are not uniquely determined by the convex set displayed. It is easy to see that:

$$y_0+/y_1/\ldots/y_m/ = \{\, y_0+\Sigma_{i>0}\beta_i y_i \mid 1 \geq \beta_1 \geq \ldots \geq \beta_m \geq 0,\ \beta_i \in k\}$$

Moreover, $y_0+/y_1/\ldots/y_m/$ is a geometric m-simplex if and only if $y_1, \ldots, y_m$ are k-linearly independent (as vectors). The <u>interior</u> of the geometric m-simplex is the set:

$$\{y_0+\Sigma_{i>0}\beta_i y_i \mid 1 > \beta_1 > \ldots > \beta_m > 0,\ \beta_i \in k\}.$$

The codimensional 1 faces of the geometric m-simplex $y_0+/y_1/\ldots/y_m/$ are:

$$y_0+y_1+/y_2/\ldots/y_m/,\ y_0+/y_1+y_2/y_3/\ldots/y_m/,\ \ldots,$$

$$y_0+/y_1/\ldots/y_{m-1}+y_m/,\ \text{and}\ y_0+/y_1/\ldots/y_{m-1}/$$

We note that no ambiguity arises from the two usages of + in the preceding display.

Suppose that $A = /y_1/\ldots/y_m/$ is a geometric m-simplex with $m \geq 2$. A can also be described by an affine fibration. The base is the geometric 2-simplex $/y_1/y_2/$ in Lsp(A). The fiber over the point $\beta_1 y_1+\beta_2 y_2$ is:

$$(\beta_1 y_1+\beta_2 y_2) + (\beta_2 \circ /y_3/\ldots/y_m/),\ 1 \geq \beta_1 \geq \beta_2 \geq 0.$$

Of course, we can use p in place of 2, $1 \leq p \leq m$. However, the case $p = 2$ is the only one we need.

We want to define the volume of a polyhedron. There are two methods. Both of these procedures are relative rather than absolute.

For the first procedure, we select a k-basis $e_1, \ldots, e_n$ for $\mathfrak{m}$. If $y_0+/y_1/\ldots/y_n/$ is a geometric n-simplex, then we have:

$$y_1 \wedge \ldots \wedge y_n = d(e_1 \wedge \ldots \wedge e_n), \ d \neq 0 \in k.$$

The oriented volume of our geometric n-simplex is defined to be $d/n!$. The (unoriented or absolute) volume is defined to be $|d|/n!$. The theory of determinant (via exterior algebra) shows that the volume is well defined. It is additive with respect to simple subdivisions and is G-invariant for an affine group G of transformations on $\mathfrak{m}$ if and only if G is a subgroup of $SA_k^{\pm}(\mathfrak{m}) = T_k(\mathfrak{m})SL_k^{\pm}(\mathfrak{m})$. For this reason, we will now assume that G is a subgroup of $SA_k^{\pm}(\mathfrak{m})$. It follows that we have a group homomorphism:

$$\text{vol} : \mathcal{P}^n(G) \longrightarrow k$$

This map has the additional properties:

$\text{vol}(\lambda \circ A) = \lambda^n \text{vol}(A)$, $\lambda \in k^+$, A any polyhedron; and

$\text{vol}(A) > 0$ for any geometric n-simplex A.

Later discussions will make it apparent that these three properties characterize volume up to a positive factor. We note that volume for affine polyhedra is essentially an algebraic concept. Geometry and/or analysis make their appearances only in the sense that the field k should be ordered.

For the second procedure, we select an ordered extension field K of k so that each element of k^+ has a (unique) square root in K^+. From the theory of ordered fields, it is known that K can be found and a minimal candidate is obtained from k by adjoining all square roots of all positive elements of k (any two ordering on the minimal candidate are Galois equivalent). We next select a positive definite inner product $< , >$ on $\mathfrak{m}$. If $A = y_0 + /y_1/\ldots/y_m/$, then its m-dimensional volume is given by:

$$\{\det(<y_i, y_j>)\}^{1/2}/m! \ (\geq 0 \text{ and lies in } K), \ 1 \leq i, j \leq m.$$

A is a geometric m-simplex if and only if its volume is positive.

When A is a geometric n-simplex, the second differs from the first by a factor of $|\det(<e_i, e_j>)|^{1/2}$. It is therefore clear that the image of the volume map is a k-vector space of dimension 1 with either definition. In the second procedure, the inner product automatically gives rise to a volume for all m-dimensional polyhedra, $0 \leq m \leq n$ (the volume of a point is taken to be 1). In this respect, the first procedure suffers because a choice of a k-basis must be made for each linear subspace of $\mathfrak{M}$ before we can define volumes on lower dimensional polyhedra. However, the economy of the second procedure is accomplished at the expense of dragging in more algebra (ordered fields and inner products). We note that when k is Archimedean, it can be identified with a subfield of the field $\mathbb{R}$ of real numbers. $\mathfrak{M}$ can be identified with the space k^n of column vectors of length n through the choice of a k-basis $e_1, \ldots, e_n$. As such, there is the usual inner product on k^n and both definitions yield the usual Jordan content. It should be noted that there does not appear to be an algebraic definition of volume for geometric m-simplices in spherical or hyperbolic spaces when $m > 2$.

There is one additional advantage of imposing a positive definite inner product. It allows cutting down the set of generators of $\wp^n(G)$. By definition, the group $\wp^n(G)$ of affine polyhedra (for the stable G-scissors congruence problem) has generators $[P]$ with P ranging over all the geometric n-simplices in $\mathfrak{M}$ and has the relators:

$[P] - [\sigma P]$, P any geometric n-simplex, $\sigma \in G$ arbitrary; and
$[P] - [Q] - [R]$, $P = Q \coprod R$ ranges over all simple subdivisions of arbitrary n-simplex P into n-simplices Q and R.

If $P = \coprod_i P_i$ is any polyhedron with P_i denoting n-simplices, we then write $[P] = \Sigma_i [P_i]$. It is not difficult to see that two polyhedra P and Q satisfy the relation $[P] = [Q]$ if and only if $P \overset{s}{\sim}_G Q$. We need a few more concepts before describing the cutting down process.

A geometric m-simplex $y_0 + /y_1/\ldots/y_m/$ is said to be <u>orthogonal</u> (with respect to the inner product $<\,,\,>$) if:

$$<y_i, y_j> = 0, \quad 1 \leq i < j \leq m.$$

It is easy to see that an orthogonal geometric m-simplex $y_0+/y_1/.../y_m/$ has exactly one other nonhomogeneous representation as an orthogonal geometric m-simplex:

$$y_0+/y_1/.../y_m/ = y_0+...+y_m+/-y_m/.../-y_1/.$$

In general, the G-transforms of an orthogonal geometric m-simplex do not have to be orthogonal. It will be so when G is a subgroup of $E(\mathfrak{m}, \langle\ ,\ \rangle) = T_k(\mathfrak{m})O_k(\mathfrak{m}, \langle\ ,\ \rangle)$.

For any subsets $A_1, ..., A_t$ of $\mathfrak{m}$, the Minkowski sum A is defined by:

$$A = A_1 + ... + A_t = \{\Sigma_i x_i \mid x_i \in A_i\}.$$

If each A_j is a geometric d(j)-simplex and $d = \Sigma_j d(j)$, then $A_1+...+A_t$ is called a <u>(basic)</u> <u>t-fold</u> <u>cylinder</u> <u>of</u> <u>dimension d</u> when we have:

each $d(j) > 0$ and $\dim_k \mathrm{Lsp}(A) = d$.

Such a cylinder is called <u>orthogonal</u> if, in addition, we have:

each A_j is an orthogonal d(j)-simplex, $d(j) > 0$, and $\mathrm{Lsp}(A_j)$ are pairwise orthogonal.

We write $A = A_1 \# ... \# A_t$ in such cases (whether it is orthogonal or not). Evidently $1 \leq t \leq n$. Unless specified, the dimension d is understood to be n and will be dropped. For $1 \leq t \leq n+1$, $\wp^n_t(G)$ is defined to be the subgroup of $\wp^n(G)$ generated by the classes of all t-fold cylinders. Decomposing cylinders into simplices, we have the decreasing <u>Minkowski</u> <u>filtration</u>:

$$\wp^n(G) = \wp^n_1(G) \supset ... \supset \wp^n_n(G) \supset \wp^n_{n+1}(G) = 0.$$

We note that a basic n-fold cylinder is just an n-dimensional parallelopiped. We note also that $[A]+[B]$, $[A+B]$ and $[A\#B]$ are generally distinct.

One of the characteristic features of the affine n-space is the existence of global similarities noted by Wallis [23; p. 2] a long time ago. It will play a decisive role in settling the affine case of the scissors congruence problem. Let $\lambda \in k^+ = \{\alpha \in k \mid \alpha > 0\}$. For any subset A of $\mathfrak{m}$, the image of A under λI_n is denoted by $\lambda \circ A = \{\lambda x \mid x \in A\}$. This is called <u>scaling</u> <u>by</u> <u>λ</u>.

We will now combine the concept of volume with simple combinatorics to describe certain decompositions of some of our polyhedra.

Let p and q be nonnegative integers with $p+q = n$. A permutation σ of $\{1, \dots, n\}$ is called a <u>(p, q)-shuffle</u> if $\sigma(i) < \sigma(j)$ holds when either $i < j \leq p$ or $p < i < j$. Similarly, σ is called a <u>(-p, q)-shuffle</u> if $\sigma(i) > \sigma(j)$ holds for $i < j \leq p$ and $\sigma(i) < \sigma(j)$ holds for $p < i < j$. For example, the identity permutation is a (p, q)-shuffle for any (p, q) and is a (-0, n)-shuffle as well as a (-1, n-1)-shuffle. For a fixed pair (p, q), the number of distinct (p, q)-shuffles as well as (-p, q)-shuffles is just $(p+q)!/p!q!$.

<u>Proposition 2.1.</u> Let $x_1, \dots, x_n$ be any k-basis of $\mathfrak{m}$. Then,

(a) $/x_1/\dots/x_p/ \# /x_{p+1}/\dots/x_n/ = \bigsqcup /x_{\sigma(1)}/\dots/x_{\sigma(n)}/$, where σ^{-1} ranges over all (p, q)-shuffles;

(b) $/x_1/ \# \dots \# /x_n/ = \bigsqcup /x_{\rho(1)}/\dots/x_{\rho(n)}/$, where ρ^{-1} ranges over all permutations; and

(c) $\bigsqcup_{i \text{ odd}} /x_i/\dots/x_1/ \# /x_{i+1}/\dots/x_n/$

$= \bigsqcup_{i \text{ even}} /x_i/\dots/x_1/ \# /x_{i+1}/\dots/x_n/$.

Moreover, if $x_1, \dots, x_n$ is an orthogonal k-basis of $\mathfrak{m}$ with respect to some specified positive definite inner product on $\mathfrak{m}$, then all the simplices and cylinders appearing in (a), (b) and (c) are orthogonal. The decompositions in (a), (b) and (c) are compatible with the actions of the full affine group of $\mathfrak{m}$. The orthogonality is compatible with the actions of the subgroup $E(\mathfrak{m}, < , >)$ of all rigid motions.

<u>Proof</u>. The interior of the geometric n-simplex $/x_{\rho(1)}/\dots/x_{\rho(n)}/$ is the set $\{ \Sigma\alpha_i x_{\rho(i)} \mid 1 > \alpha_1 > \dots > \alpha_n > 0,\ \alpha_i \in k\}$. In terms of the given ordered basis $x_1, \dots, x_n$, each interior point has coordinates: $(\beta_1, \dots, \beta_n)$, where $\beta_i = \alpha_{\rho^{-1}(i)}$. It is clear that the permutation ρ is uniquely determined by each interior point. This quickly shows that the right hand simplices of (a) (respectively (b)) are pairwise interior disjoint; it also shows that the right hand sides are contained in the respective left hand sides in (a) and (b). A volume calculation then yields the equality in (a) and (b).

Let σ denote a $(-p, q)$-shuffle. Suppose that $\sigma(p) < \sigma(p+1)$, then σ is also a $(-(p-1), q+1)$-shuffle. Suppose that $\sigma(p) > \sigma(p+1)$, then σ is also a $(-(p+1), q-1)$-shuffle. These are clearly the only possible representations of σ as a $(-r, s)$-shuffle. The ambiguity is removed by the specification of the parity of r. Assertion (c) then follows from (a). Q.E.D.

Let $P = \mathrm{ccl}\{x_0, \ldots, x_m\}$ be any geometric m-simplex, $m > 0$. A (proper) simple subdivision of P is defined by the specification of a point y on the interior of any 1-face of P. Namely, $y = \alpha x_i + \beta x_j$, $i \neq j$, $\alpha + \beta = 1$ and $\alpha, \beta \in k^+$. We then have $P = P' \coprod P''$ with:

$$P' = \mathrm{ccl}\{x_0, \ldots, x_{i-1}, \hat{x}_i, y, x_{i+1}, \ldots, x_n\}; \text{ and}$$

$$P'' = \mathrm{ccl}\{x_0, \ldots, x_{j-1}, \hat{x}_j, y, x_{j+1}, \ldots, x_n\}.$$

If either $i = j$ or $\alpha\beta = 0$, the subdivision is improper and one of P', P'' must be left out. If $i \neq j$ and we allow $\alpha\beta < 0$, then we can move one of P', P'' to the left side and speak of <u>superdivision</u>. We note that P is not changed under arbitrary permutation of its vertices so we can always take $i = j-1$. In the nonhomogeneous formulation, we write $P = y_0 + /y_1/\ldots/y_m/$. If we specify αy_j, $0 < \alpha < 1$, $\alpha \in k$, $1 \leq j \leq m$ and set $\beta = 1-\alpha$, then a simple subdivision at αy_j yields $P = P' \coprod P''$ with:

$$P' = y_0 + /y_1/\ldots/y_{j-1}/\alpha y_j/\beta y_j + y_{j+1}/\ldots/y_m/; \text{ and}$$

$$P'' = y_0 + \alpha y_1 + /\beta y_1/y_2/\ldots/y_m/, \text{ when } j = 1,$$

$$= y_0 + /y_1/\ldots/y_{j-2}/y_{j-1} + \alpha y_j/\beta y_j/y_{j+1}/\ldots/y_m/, \text{ when } j > 1.$$

We note that P has $(m+1)!$ distinct nonhomogeneous representations. The preceding simple subdivision may have a messier description with respect to a different nonhomogeneous representation of P. We also note that a simple subdivision of an orthogonal geometric m-simplex need not yield orthogonal simplices. This is a feature that prevents us from dealing exclusively with orthogonal simplices. In spite of this, the orthogonal geometric n-simplices will play an important role.

Proposition 2.2. Let $e_1, \ldots, e_n$ be a k-basis for $\mathfrak{m}$. Let $x_0, \ldots, x_{n+1} \in \mathfrak{m}$. For $0 \leq i \leq n+1$, let $A_i = \mathrm{ccl}\{x_0, \ldots, \hat{x}_i, \ldots, x_{n+1}\}$ and let $d_i \in k$ so that:

$$(x_1-x_0)\wedge\ldots\wedge(x_{i-1}-x_{i-2})\wedge(x_{i+1}-x_{i-1})\wedge(x_{i+2}-x_{i+1})\wedge\ldots\wedge(x_{n+1}-x_n)$$
$$= d_i(e_1\wedge\ldots\wedge e_n).$$

Define A_i^+ (resp. A_i^-) to be A_i if $(-1)^i d_i > 0$ (resp. < 0) and to be $\emptyset$ if we have $(-1)^i d_i \leq 0$ (resp. ≥ 0). Then, $\coprod_i A_i^+ = \coprod_i A_i^-$.

Proof. A_i is a geometric n-simplex if and only if $d_i \neq 0$. Indeed, $d_i/n!$ is the oriented volume of the (possibly degenerate) n-simplex A_i. We can therefore assume that some A_i is a geometric n-simplex. If we exchange adjacent vertices x_{i-1}, x_i, then the sides of the asserted equation are exchanged. We note also that our assertion is trivial if $x_i = x_j$ for some $i \neq j$. We can therefore assume that $x_0, \ldots, x_{n+1}$ are n+2 distinct points of $\mathfrak{m}$. There is then at least one affine hyperplane: $H = \mathrm{Asp}(x_1, \ldots, x_n)$ for example, separating x_0 and x_n. By changing coordinates, we may assume $H \cap \mathrm{ccl}\{x_0, x_n\}$ to be the origin of $\mathfrak{m}$. The set $A = \mathrm{ccl}(x_0, \ldots, x_{n+1})$ is now the interior disjoint union of two cones with apices x_0, x_{n+1} and with common base $\mathrm{ccl}(0, x_1, \ldots, x_n) \subset H$. We may proceed by induction on n. If $j \neq 0$, n+1 and $d_j \neq 0$, then A_j is subdivided at 0. The definitions of A_j^+ and A_j^- are such that the induction hypotheses can be checked easily. We omit further details and note only that there are several possibilities even when n = 2. Q. E. D.

Corollary 2.3. $\mathcal{P}^n(G)$ is generated by the set of all [A] with A ranging over all geometric n-simplices of the form $/y_1/\ldots/y_n/$.

Proof. It is enough to consider the generation of a random geometric n-simplex $\mathrm{ccl}\{x_0, \ldots, x_n\}$. The assertion follows from Proposition 2.2 with $x_{n+1} = 0$. Q. E. D.

Let $\mathfrak{m}$ be given a positive definite inner product $<\ ,\ >$. Let $x \in \mathfrak{m}$ and let H be an affine hyperplane of dimension n-1. Since the normal component of a vector with respect to a codimensional 1 subspace of $\mathfrak{m}$ lies in $\mathfrak{m}$, we can find the foot of the perpendicular from x to H.

Theorem 2.4. Let $\mathfrak{M}$ be given a positive definite inner product $< , >$. Let L be a 1-dimensional subspace with orthogonal complement H. Assume the group G contains the subgroup $T_k(\mathfrak{M})$ of all translations. Then $\mathcal{P}^n(G)$ can be generated by the set of all [A] with A ranging over all orthogonal geometric n-simplices of the form $/y_1/\ldots/y_n/$ where $y_1 \in L$. (In general, $y_j \in H$ for $j > 1$. However, if $n > 2$, we may not prescribe ky_2.)

Proof. It suffices to show the generation of a geometric n-simplex $ccl(x_0, \ldots, x_n)$ by the orthogonal n-simplices described. By the assumption on G, we may translate at will. By using coordinates normal to H, we may speak of the H-height of the vertices x_i, $0 \leq i \leq n$.

Suppose there are 4 distinct vertices x_a, x_b, x_c and x_d so that x_a and x_b have the same H-height while x_c and x_d have the same H-height distinct from that of x_a, x_b. Since k is necessarily infinite, we can subdivide our simplex at an interior point z of $ccl(x_a, x_c)$ so that z has H-height distinct from the H-height of any of the vertices x_i. Applying Proposition 2.2 with $x_{n+1} = z$ we can assume that $ccl(x_0, \ldots, x_n)$ has the property:

(*) At most one H-height contains more than 1 vertex.

The H-height with more than 1 vertex will be called exceptional. When it does not occur, we arbitrarily designate the least H-height with 1 vertex as exceptional. We now have two cases:

Case 1. There is only one vertex, say x_0, with nonexceptional H-height. We then have $H = Lsp(x_1, \ldots, x_n)$. Let x_{n+1} be the foot of the perpendicular from x_0 to its opposite face. Applying Proposition 2.2 together with translations, $[ccl(x_0, \ldots, x_n)]$ can be generated by [A] with A having the form $/y_1/\ldots/y_n/$ with $y_1 \in L$ and $y_2, \ldots, y_n \in H$. As it stands, the n-simplex $/y_1/\ldots/y_n/$ is not necessarily orthogonal when $n > 2$. It is a cone with apex at 0 and with base $y_1 + /y_2/\ldots/y_n/$. In the base (n-1)-simplex, we drop a perpendicular from the vertex y_1 to the opposite face in $Asp(y_1 + y_2 + /y_3/\ldots/y_n/)$. Applying Proposition 2.2 again, $[/y_1/\ldots/y_n/]$ can be generated in $\mathcal{P}^n(G)$ by [B] with B having the form $/y_1/z_2/\ldots/z_n/$ where $z_j \in H$ and $< z_2, z_j > = 0$ for $j > 2$. This process can be repeated with the face $y_1+z_2+/z_3/\ldots/z_n/$. In place of the cone construction, we use the geometric join. The desired generation is obtained by induction.

Case 2. There are at least two vertices, say x_a and x_b, with nonexceptional H-heights. Let x be the point of intersection of $Asp(x_a, x_b)$ with the affine hyperplane formed by all points having the exceptional H-height. We subdivide or superdivide our given simplex at x. The two resulting simplices will each have one less vertex with nonexceptional H-height than the one given. In a finite number of steps, the Case 2 is reduced to Case 1 already considered. Q.E.D.

Theorem 2.5. Let volume be defined with respect to a k-basis $e_1, \ldots, e_n$ of $\mathfrak{m}$. Suppose that G is a subgroup between $T_k(\mathfrak{m})$ and $SA_k^{\pm}(\mathfrak{m})$. Then the volume map is a bijection between $\mathcal{P}_n^n(G)$ and k.

Proof. When n = 1, each element of $\mathcal{P}^1(G)$ is represented as $\pm[/\lambda e_1/]$ with $\lambda \geq 0$ and its volume is $\pm\lambda$. Our assertion is therefore clear. Suppose $n > 1$. We assert that each [A], A a basic n-fold cylinder, is equal to $[/\lambda e_1/\#\ldots\#/e_n/]$ where λ is the absolute volume of A. By working with 2 coordinates at a time, we can reduce the general case to n = 2. More precisely, SL(n;k) is generated by its transvections σ. Namely, we can find $i \neq j$ so that $\sigma(e_s) = e_s$ for $s \neq i$ and $\sigma(e_i) = e_i + \alpha e_j$ for some $\alpha \in k$. When n = 2, the pictures below show that $[/x/\#/y/] = [/x/\#/\alpha x + y/]$ in $\mathcal{P}_2^2(G)$. Our assertion then follows from determinant theory. Q.E.D.

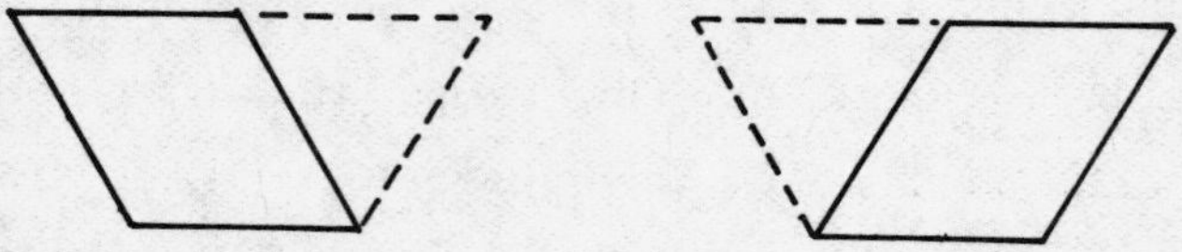

Proposition 2.6. Each element of the group $\mathcal{P}_t^n(G)/\mathcal{P}_n^n(G)$ can be represented by a finite sum (no subtraction needed) $\Sigma_i[A_i]$ where A_i is a suitable basic t-fold cylinder.

Proof. Let B be any basic t-fold cylinder. Combining (a) and (b) of Proposition 2.1, we can find a finite number of t-fold cylinders B_j

depending only on B so that $B \coprod (\coprod_j B_j)$ is an n-fold cylinder. As a consequence, the coset of $-[B]$ can be replaced by the coset of $\Sigma_j [B_j]$. This yields the desired assertion. Q. E. D.

Corollary 2.7. Let P be any n-fold cylinder. Let $A = \coprod A_i$ be contained in P and write $P = A \coprod B$. If each A_i is a basic t-fold cylinder, $1 \le t \le n$, then $[B] \in \mathcal{P}_t^n(G)$.

We leave the proof to the patient reader. We note that the conclusion does not say that B has a representation $\coprod_j B_j$ where each B_j is a basic t-fold cylinder.

3. CANONICAL DECOMPOSITIONS.

We record some fundamental results due to Sydler and Hadwiger. Except for some minor differences, these results can be found in Hadwiger's book [31; Chapters 1 and 2].

Theorem 3.1. (First Canonical Decomposition) Let $x_1, \ldots, x_n$ be any k-basis of $\mathfrak{m}$ and let $\alpha, \beta \in k^+$. Then:

$$(\alpha + \beta) \circ /x_1/\ldots/x_n/ = \alpha \circ /x_1/\ldots/x_n/ \coprod \ldots$$
$$\coprod (\alpha(x_1 + \ldots + x_i)) + \{(\beta \circ /x_1/\ldots/x_i/) \# (\alpha \circ /x_{i+1}/\ldots/x_n/)\} \coprod \ldots$$
$$\coprod (\alpha(x_1 + \ldots + x_n)) + \{(\beta \circ /x_1/\ldots/x_n/\}.$$

If $/x_1/\ldots/x_n/$ is orthogonal with respect to some specified positive inner product $\langle\, ,\, \rangle$ on $\mathfrak{m}$, then the simplices and cylinders displayed are all orthogonal. Except for the first and the last term, all other terms on the right side of the equation displayed are basic 2-fold cylinders.

The proof is not difficult and can be found in [31; p. 18-19]. The following pictures for n = 2 and 3 are indicative of the proof.

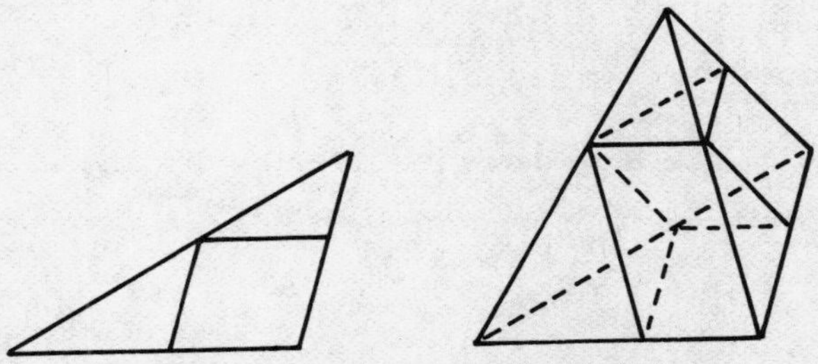

Theorem 3.2. (Second Canonical Decomposition) Let $x_1, \ldots, x_n$ be any k-basis of $\mathfrak{m}$. For $1 \leq i \leq j \leq n$, let $A(i,j) = /x_i/\ldots/x_j/$. Let $[A(i)]$ lie in $\mathcal{P}^n(T_k(\mathfrak{m}))$ be defined by:

$[A(i)] = \Sigma\,[A(1,j_1) \# \ldots \# A(j_{i-1}+1,n)]$, where the sum extends over all sequences of integers: $0 = j_0 < j_1 < \ldots < j_i = n$.

Let m be any positive integer, then:

$$[m\,oA(1,n)] = \Sigma_{1 \leq i \leq n} \binom{m}{i} [A(i)] \text{ in } \mathcal{P}^n(T_k(\mathfrak{m})).$$

The integer $\binom{m}{i}$ is the evaluation of the polynomial $\binom{X}{i} = \frac{X(X-1)\ldots(X-i+1)}{i!}$ at $X = m$. The polynomial lies in $\mathbb{Q}[X]$, has degree i with leading coefficient $1/i!$ and constant term 0 when $1 \leq i \leq n$. Moreover, if $/x_1/\ldots/x_n/$ is an orthogonal simplex with respect to some positive definite inner product $\langle\ ,\ \rangle$ on $\mathfrak{m}$, then each $A(i)$ may be realized as a finite interior disjoint union of orthogonal i-fold cylinders.

The assertions follow from the first canonical decomposition together with an induction on m. The details can be found in [31; p. 32]. It should be noted that the precise translations in the first canonical decomposition and the elementary properties of $\binom{X}{i}$ in the second canonical decomposition both play a role later.

Remark 3.3. We already observed that a G-invariant volume function exists for any subgroup G of $SA_k^{+}(\mathfrak{m})$. As a consequence, condition (VP) in Zylev's Theorem is valid for the G-scissors congruence data on $\mathfrak{m}$ when G is contained in $SA_k^{+}(\mathfrak{m})$.

Theorem 3.4. (Jordan Approximation Theorem) Let k be an Archimedean ordered field so that k can be identified with a subfield of $\mathbb{R}$. Let A be any polyhedron in $\mathfrak{m}$ with vol $A > 0$. Let $\epsilon \in \mathbb{Q}^+$. Then there exist polyhedra A_{in} and A_{out} with the following properties:

(a) A_{in} and A_{out} are both pairwise interior disjoint unions of suitable finite number of n-fold cylinders, $A_{in} \subset A \subset A_{out}$; and

(b) $\text{vol}\,A_{in} < \text{vol}\,A < \text{vol}\,A_{out}$, $0 < \text{vol}\,A_{out} - \text{vol}\,A_{in} < \epsilon$.

Proof. We will only sketch the derivation of this wellknown result from the second canonical decomposition. A can be taken to be a geometric n-simplex. For $\lambda \in k^+$, we have $A = \lambda \circ \lambda^{-1} \circ A$. We apply the second canonical decomposition to $B = \lambda^{-1} \circ A$ with λ small and m large (here we need k to be Archimedean) and calculate volume. With appropriate λ and m, we can take $A_{in} = \lambda \circ (B(n))$. We note further that A is the interior disjoint union of A_{in} with a large number of geometric n-simplices each with small volume. For each of these small n-simplices, we apply (b) of Proposition 2.1 and add on enough small n-simplices to make up a small parallelopiped. Add on all these small parallelopipeds to A_{in} and perform a volume calculation show that we have the desired A_{out} when λ and m are selected appropriately. Q.E.D.

Theorem 3.5. Let $n \geq 2$. Identify $\mathfrak{M}$ with k^n through a k-basis $e_1, \ldots, e_n$ of $\mathfrak{M}$ and give k^n the usual inner product $< , >$. Let G be any subgroup between $T_k(\mathfrak{M}) = T(n;k)$ and $E_k(\mathfrak{M}, < , >) = E(n;k)$. Equivalent statements are

(a) k is Archimedean;

(b) condition (VA) of Zylev's Theorem holds; and

(c) stable G-scissors congruence on $\mathfrak{M}$ is equivalent with G-scissors congruence on $\mathfrak{M}$.

Proof. (a) implies (b). Define volume as before with respect to the k-basis $e_1, \ldots, e_n$ of $\mathfrak{M}$. Using (a), condition (VA) of Zylev's Theorem is a consequence of the following assertion:

(*) if A and B are polyhedra with vol A < vol B, then there is a polyhedron C with $C \subset B$ and C is G-scissors congruent to A.

Using Jordan approximation, A and B can be replaced by A_{out} and B_{in} respectively. The desired conclusion would follow if we can strengthen Theorem 2.5 to the following assertion:

(**) if k is Archimedean, then every n-fold cylinder A is $T_k(\mathfrak{M})$-scissors congruent to $/\lambda e_1/\ldots/e_n/$, $\lambda = \text{vol } A$.

We note that (**) is clear when n = 1.

As in the proof of Theorem 2.5, the argument reduces to $n = 2$. The pictorial argument for $n = 2$ needs to be replaced by the following pictures:

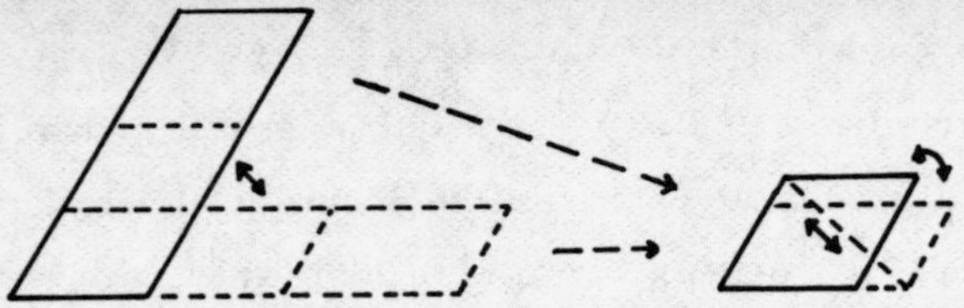

We note that the preceding pictures yield a proof of the following (much more complicated) result in the case $n = 2$.

(***) Let $P \# Q$ be a basic 2-fold cylinder in $\mathfrak{M}$. For any $\lambda \in k^+$,

$$[(\lambda \circ P) \# Q] = [P \# (\lambda \circ Q)] \text{ in } \mathcal{P}_2^n(T_k(\mathfrak{M})) \bmod \mathcal{P}_3^n(T_k(\mathfrak{M})).$$

As it stands, (***) is a stable result. However, when combined with the present theorem, it will ultimately lead to a solution of the G-scissors congruence problem for the affine case over an Archimedean field.

(b) implies (c). We have already noted that condition (VP) follows from the existence of an G-invariant volume. The present implication is just the content of Zylev's Theorem. Indeed, the present implication is valid for any subgroup G of $SA_k^{\pm}(\mathfrak{M})$.

(c) implies (a). This implication uses all the assumptions. The proof is due to Hilbert and proceeds by contradiction. Assume (c) holds but (a) fails. We can find α, β in k^+ so that $N\alpha < \beta$ holds for every positive integer N. Dividing by α, we may assume that $N < \beta$ holds for every positive integer N. By assumption, we have a G-invariant distance as well as volume on $\mathfrak{M}$ with values in a suitable ordered field extension K of k. Let $A = /e_1/\#\ldots\#/e_n/$ and $B = /\beta e_1 + e_2/\#/e_2/\#\ldots\#/e_n/$. According to Theorem 2.5, A and B are stably G-scissors congruent. By assumption (c), they must be G-scissors congruent. We can therefore find decompositions $A = \coprod_i A_i$, $B = \coprod_i B_i$, $1 \leq i \leq t$, so that $B_i = \sigma_i A_i$ holds for suitable σ_i in G. By triangle inequality, the distance between any two points in A is at most n. One of the edges of B has length d with $d^2 = 1 + \beta^2$. It follows that at least one of the B_i's must have an edge of length h with $h^2 \geq (1+\beta^2)/t^2$. Since σ_i preserves distance, the corresponding edge in A_i must have length h and satisfies the inequality: $n^2 \geq h^2$. We therefore have $t^2n^2 - 1 \geq \beta^2$. Since t and n are positive integers, this inequality contradicts the fact that $N < \beta$ holds for every positive integer N. Q.E.D.

Proposition 3.6. Let G be any subgroup between $T_k(\mathfrak{m})$ and $SA^{\pm}_k(\mathfrak{m})$. Then:

(a) $\mathcal{P}^n_i(G)$ is stable with respect to scaling operations; and

(b) if $[A] \in \mathcal{P}^n_i(G)$ and $m \in \mathbb{Z}^+$, then $[m \circ A] \equiv m^i[A] \bmod \mathcal{P}^n_{i+1}(G)$.

Proof. (a) follows from the fact that scaling is distributive with respect to Minkowski sum. For (b), we use the first canonical decomposition. The dependence on translation is gotten rid of by the assumption that G contains $T_k(\mathfrak{m})$. Scaling is distributive with respect to interior disjoint union so that we may assume A to be a basic i-fold cylinder. The desired result follows from the distributive law of scaling with respect to Minkowski sum together with the first canonical decomposition applied to each factor of A. Q.E.D.

Theorem 3.7. Let G be any subgroup between $T_k(\mathfrak{m})$ and $SA^{\pm}_k(\mathfrak{m})$. Let $1 \leq i \leq n$. The Minkowski filtration is a decreasing filtration by $\mathbb{Q}$-vector subspaces $\mathcal{P}^n_i(G)$. Scaling by $m \in \mathbb{Q}^+$ induces scalar multiplication by m^i on the $\mathbb{Q}$-vector space $\mathcal{P}^n_i(G)/\mathcal{P}^n_{i+1}(G)$ where $\mathcal{P}^n_{n+1}(G) = 0$.

Proof. Theorem 2.5 shows that $\mathcal{P}^n_n(G)$ is isomorphic to k as an additive group so that it is a $\mathbb{Q}$-vector space. Proposition 3.6 together with a downward induction on n imply that each $\mathcal{P}^n_i(G)$ is infinitely divisible as an additive group. In view of (b) of Proposition 3.6, we only need to show that $\mathcal{P}^n_i(G)/\mathcal{P}^n_{i+1}(G)$ is torsionfree, $1 \leq i < n$. According to Proposition 2.6, each element of $\mathcal{P}^n_i(G)/\mathcal{P}^n_{i+1}(G)$ can be represented as a finite sum $\Sigma_j[A_j]$ where A_j represents a basic i-fold cylinder. If $m(\Sigma_j[A_j]) \in \mathcal{P}^n_{i+1}(G)$ holds for some positive integer m, then (b) of Proposition 3.6 shows that $\Sigma_j[m \circ A_j] \in \mathcal{P}^n_{i+1}(G)$. We always have $m^{-1} \circ m \circ A_j = A_j$. We conclude from (a) of Proposition 3.6 that $\Sigma_j[A_j]$ must lie in $\mathcal{P}^n_{i+1}(G)$ so that $\mathcal{P}^n_i(G)/\mathcal{P}^n_{i+1}(G)$ is torsionfree. Q.E.D.

4. RATIONAL FUNCTIONALS.

We have seen that scaling by elements of k^+ defines automorphisms of $\mathcal{P}^n(G)$ and stablizes the filtration $\mathcal{P}_i^n(G)$ so that $\mathcal{P}^n(G)$ can be viewed as $\mathbb{Q}$-representation space for the group k^+. Let Y be any $\mathbb{Q}$- (resp. k-) vector space. We view Y as a $\mathbb{Q}$- (resp. k-) representation space for the multiplicative group $\mathbb{Q}^+$ (resp. k^+) via the obvious embedding of $\mathbb{Q}^+$ (resp. k^+) in $\mathbb{Q}$ (resp. k). An element $\chi \in \mathrm{Hom}_{\mathbb{Z}}(\mathcal{P}^n(G), Y)$ is said to be of $\underline{\mathbb{Q}^+}$ (resp. $\underline{k^+}$) weight i if:

$$\chi[\lambda \circ A] = \lambda^i \chi[A] \text{ holds for all } \lambda \in \mathbb{Q}^+ \text{ (resp. } k^+\text{) and all } [A] \in \mathcal{P}^n(G).$$

It is immediate that such an element χ is completely determined by the following properties:

(RF_1) χ is a Y-valued function of geometric n-simplices;

(RF_2) χ is additive with respect to simple subdivisions of geometric n-simplices;

(RF_3) χ is G-invariant, i.e., $\chi(\sigma(A)) = \chi(A)$ holds for σ in G and arbitrary geometric n-simplex A; and

(RF_4) $\chi(\lambda \circ A) = \lambda^i \chi(A)$ holds for $\lambda \in \mathbb{Q}^+$(resp. k^+) and arbitrary geometric n-simplex A.

Properties (RF_1), (RF_2) and (RF_3) and the functorial properties of $\mathcal{P}^n(G)$ characterize elements of $\mathrm{Hom}_{\mathbb{Z}}(\mathcal{P}^n(G), Y)$. Property ($RF_4$) is the usual definition of weight i. Since Y and $\mathcal{P}^n(G)$ are vector spaces over the quotient field $\mathbb{Q}$ of $\mathbb{Z}$, we have $\mathrm{Hom}_{\mathbb{Z}}(\mathcal{P}^n(G), Y) = \mathrm{Hom}_{\mathbb{Q}}(\mathcal{P}^n(G), Y)$. Clearly, if χ has k^+ weight i, then it has $\mathbb{Q}^+$ weight i. The converse is usually not valid. The set of all elements of weight i is a subspace of $\mathrm{Hom}_{\mathbb{Z}}(\mathcal{P}^n(G), Y)$ where the vector space structure (over $\mathbb{Q}$, resp. k) comes from that of Y. As it stands, $\mathcal{P}^n(G)$ only has a $\mathbb{Q}$-vector space structure. It will ultimately be given a k-vector space structure (actually quite messy to describe) so that scalar multiplication by $\lambda \in k^+$ on the quotient space $\mathcal{P}_i^n(G)/\mathcal{P}_{i+1}^n(G)$ will correspond to scaling a polyhedron by $\lambda^{1/i}$ (provided that $\lambda^{1/i} \in k^+$).

We can and will identify $\mathrm{Hom}_{\mathbb{Z}}(\mathcal{P}^n(G), Y)$ with the space of $G/T_k(\mathfrak{m})$ invariants on $\mathrm{Hom}_{\mathbb{Z}}(\mathcal{P}^n(T_k(\mathfrak{m})), Y)$ with Y viewed as a trivial G-module and with $G/T_k(\mathfrak{m})$ acting on $\mathcal{P}^n(T_k(\mathfrak{m}))$ through the action of G on the set of all geometric n-simplices. We therefore have the identification:

$$\mathrm{Hom}_{\mathbb{Z}}(\mathcal{P}^n(G), Y) \cong H^0(G/T_k(\mathfrak{m}), \mathrm{Hom}_{\mathbb{Z}}(\mathcal{P}^n(T_k(\mathfrak{m})), Y))$$

where $G/T_k(\mathfrak{m})$ is viewed as a discrete group and the cohomology is the Eilenberg-MacLane cohomology discrete groups. We will return to this point in Chapter 5 when we study the syzygy problem.

The weight spaces for distinct weights i are independent as subspaces of $\mathrm{Hom}_{\mathbb{Z}}(\mathcal{P}^n(T_k(\mathfrak{m})), Y)$. They are stable with respect to $G/T_k(\mathfrak{m})$ because $\{\lambda I_n \mid \lambda \in k^+\} \cong k^+$ is contained in the center of $A_k(\mathfrak{m})/T_k(\mathfrak{m}) \cong GL(n;k)$. The elements of $\mathrm{Hom}_{\mathbb{Z}}(\mathcal{P}^n(G), Y)$ are called <u>Y-valued G-invariant functionals on polyhedra</u>, or simply <u>functionals</u>. We will call them <u>rational functionals</u> when we want to emphasize the fact that they are only $\mathbb{Q}$-vector space homomorphisms. The simplest example is volume and it has k-weight n.

In analogy with E. Artin's axiomatic definition of determinant, we will give an alternate description of the elements of $\mathrm{Hom}_{\mathbb{Z}}(\mathcal{P}^n(T_k(\mathfrak{m})), Y)$. Let $GL_k(\mathfrak{m})$ denote the set of all ordered basis, $x = (x_1, \ldots, x_n)$, of the k-vector space $\mathfrak{m}$. Let $1 \leq i \leq n$ and let $\alpha, \beta \in k^+$ with $\alpha + \beta = 1$. The following two bases of $\mathfrak{m}$ are said to result from x by a <u>simple subdivision at</u> αx_i :

$$y = (x_1, \ldots, x_{i-1}, \alpha x_i, \beta x_i + x_{i+1}, x_{i+2}, \ldots, x_n),$$
$$z = (x_1, \ldots, x_{i-2}, x_{i-1} + \alpha x_i, \beta x_i, x_{i+1}, \ldots, x_n).$$

<u>Proposition 4.1</u>. With the preceding notations, there is a natural bijection between $\mathrm{Hom}_{\mathbb{Z}}(\mathcal{P}^n(T_k(\mathfrak{m})), Y)$ and the subset of all maps $\chi : GL_k(\mathfrak{m}) \longrightarrow Y$ having the following properties:

(a) if $x = (x_1, \ldots, x_n)$, $y = (y_1, \ldots, y_n) \in GL_k(\mathfrak{m})$ with $/x_1/\ldots/x_n/$ $T_k(\mathfrak{m})$-congruent to $/y_1/\ldots/y_n/$, then $\chi(x) = \chi(y)$; and

(b) if $y, z \in GL_k(\mathfrak{m})$ result from $x \in GL_k(\mathfrak{m})$ by a simple subdivision, then $\chi(x) = \chi(y) + \chi(z)$, i.e., χ is additive under subdivision.

<u>Proof</u>. Let $\chi \in \mathrm{Hom}_{\mathbb{Z}}(\mathcal{P}^n(T_k(\mathfrak{m})), Y)$ and define $c(\chi)$: $GL_k(\mathfrak{m}) \longrightarrow Y$ by:

$$c(\chi)(x_1, \ldots, x_n) = \chi[/x_1/\ldots/x_n/], \quad x = (x_1, \ldots, x_n) \in GL_k(\mathfrak{m})$$

We immediately have (a) and (b). Conversely, let $\chi : GL_k(\mathfrak{m}) \longrightarrow Y$ satisfy (a) and (b). We want to find $c'(\chi)$ in $\mathrm{Hom}_{\mathbb{Z}}(\mathcal{P}^n(T_k(\mathfrak{m})), Y)$ so that:

$$c'(\chi)[/x_1/\ldots/x_n/] = \chi(x_1,\ldots,x_n),\ (x_1,\ldots,x_n) \in GL_k(\mathfrak{m})$$

$c'(\chi)$ is unique because $\mathcal{P}^n(T_k(\mathfrak{m}))$ is generated by all $[/x_1/\ldots/x_n/]$. We need to show the existence of $c'(\chi)$. Suppose that $ccl(z_0,\ldots,z_n)$ is any geometric n-simplex in $\mathfrak{m}$. There are $(n+1)!$ ways of representing this geometric as $z + /x_1/\ldots/x_n/$. Select one such representation and let $c'(\chi)(ccl(z_0,\ldots,z_n))$ denote $\chi(x_1,\ldots,x_n)$. Condition (a) guarantees that $c'(\chi)$ is a well defined Y-valued function on geometric n-simplices in $\mathfrak{m}$. It is clear that $c'(\chi)$ is $T_k(\mathfrak{m})$ invariant. Conditions (a) and (b) together imply that $c'(\chi)$ is additive with respect to simple subdivision of geometric n-simplices. Since Y is a $\mathbb{Q}$-vector space, the functorial properties of $\mathcal{P}^n(T_k(\mathfrak{m}))$ allow us to extend $c'(\chi)$ in a unique way to an element (also denoted by $c'(\chi)$) of $\mathrm{Hom}_{\mathbb{Z}}(\mathcal{P}^n(T_k(\mathfrak{m})), Y)$. This concludes the proof of existence. By checking on generators, c and c' are then bijections inverse to each other. Q. E. D.

<u>Remark 4.2.</u> With modification on (a), the preceding result extends to any subgroup G between $T_k(\mathfrak{m})$ and $A_k(\mathfrak{m})$. In order to retain the example of volume, we need to assume that G is contained in $SA_k^{\pm}(\mathfrak{m})$.

<u>Proposition 4.3.</u> Let $A = A(1,n) = /x_1/\ldots/x_n/$ be any geometric n-simplex in $\mathfrak{m}$. Let $[A(i)] \in \mathcal{P}_i^n(T_k(\mathfrak{m}))$ be associated to A as in Theorem 3.2. Let $\chi_j \in \mathrm{Hom}_{\mathbb{Z}}(\mathcal{P}^n(T_k(\mathfrak{m})), Y)$ be of $\mathbb{Q}^+$-weight j. Then:

(a) $\chi_j[A(i)] = 0$ holds for $i > j$;

(b) $\chi_j[A] = \chi_j[A(j)]/j!$; and

(c) $\chi_j = 0$ if and only if χ_j vanishes on $\mathcal{P}_j^n(T_k(\mathfrak{m}))$.

<u>Proof.</u> According to Theorem 3.2, we have:

$$m^j\chi_j[A] = \Sigma_{1\le i\le n}\binom{m}{i}\chi_j[A(i)],\ m \in \mathbb{Z}^+$$

According to (b) of Proposition 3.6, we have:

$$m^j \chi_j[A(i)] \equiv m^i \chi_j[A(i)] \bmod \chi_j(\mathcal{P}^n_{i+1}(T_k(\mathfrak{m}))),\ m \in \mathbb{Z}^+.$$

By induction on i from n+1 down to i = j+1, assertion (a) follows from the second of our formulae. Feeding this information into the first formula, we can restrict the summation to the range $1 \leq i \leq j$. We now recall the assertions in Theorem 3.2 about the polynomial $\binom{X}{i}$. Such a polynomial is uniquely determined by its values at i+1 distinct points of $\mathbb{Q}$. Since $\chi_j[A]$ and $\chi_j[A(i)]$ lie in a finite dimensional $\mathbb{Q}$-subspace of Y independent of m, it is immediate that $\chi_j[A]$ must be equal to the coefficient $\chi_j[A(j)]/j!$ of m^j in our first formula. This is just assertion (b). If χ_j vanishes on $\mathcal{P}^n_j(T_k(\mathfrak{m}))$, then (b) shows that χ_j is 0. This yields (c). Q.E.D.

Proposition 4.4. Let Y be any $\mathbb{Q}$-vector space and let G be any subgroup between $T_k(\mathfrak{m})$ and $A_k(\mathfrak{m})$. Then:

(a) $\mathrm{Hom}_{\mathbb{Z}}(\mathcal{P}^n_i(G)/\mathcal{P}^n_{i+1}(G), Y)$ is canonically isomorphic to the $\mathbb{Q}$-vector space of Y-valued functions of $\mathbb{Q}^+$-weight i on $\mathcal{P}^n(G)$;

(b) $\mathrm{Hom}_{\mathbb{Z}}(\mathcal{P}^n(G), Y)$ is isomorphic to the direct sum of the $\mathbb{Q}$-vector spaces $\mathrm{Hom}_{\mathbb{Z}}(\mathcal{P}^n_i(G)/\mathcal{P}^n_{i+1}(G), Y)$, $1 \leq i \leq n$.

Proof. Let χ have $\mathbb{Q}^+$-weight i in $\mathrm{Hom}_{\mathbb{Z}}(\mathcal{P}^n(G), Y)$. Proposition 4.3 shows that χ can be viewed as an element of $\mathrm{Hom}_{\mathbb{Z}}(\mathcal{P}^n_i(G)/\mathcal{P}^n_{i+1}(G), Y)$ and that this defines an injective map of $\mathbb{Q}$-vector spaces.

We next show that every element $\chi \in \mathrm{Hom}_{\mathbb{Z}}(\mathcal{P}^n(G), Y)$ can be decomposed as $\Sigma_{1\leq i\leq n}\chi_i$ where χ_i has $\mathbb{Q}^+$-weight i. From Theorem 3.2, we have:

$$\chi[m \circ A] = \Sigma_{1\leq i\leq n}\binom{m}{i}\chi[A(i)],\ m \in \mathbb{Z}^+.$$

Since $\binom{X}{i}$ is a polynomial in $\mathbb{Q}[X]$ of degree i and with constant term 0 when $1 \leq i \leq n$, we can find $\chi_i(A)$ in Y so that we have a formal identity:

$$\Sigma_{1\leq i\leq n}\binom{X}{i}\chi[A(i)] = \Sigma_{1\leq i\leq n} X^i \chi_i(A) \ \text{ in } \mathbb{Q}[X]\otimes_{\mathbb{Q}} Y$$

Specializing X to $m \in \mathbb{Z}^+$ leads us to the formula:

$$\chi[m \circ A] = \Sigma_{1\leq i\leq n}\chi_i(A)\, m^i,\ m \in \mathbb{Z}.$$

The distributivity of scaling with respect to interior disjoint union together with the additivity of χ with respect to simple subdivision imply the additivity of χ_i with respect to simple subdivision. Similarly, the G-invariance of χ implies the G-invariance of each χ_i. If we let $A = p \circ B$ with $p \in \mathbb{Z}^+$ and use $(mp) \circ B = m \circ (p \circ B)$, then $\chi_i(p \circ B) = p^i \chi_i(B)$ holds so that χ_i has $\mathbb{Q}^+$-weight i. Thus we have shown the existence of the desired decomposition: $\chi = \Sigma_{1 \leq i \leq n} \chi_i$. In all these assertions, we have repeatedly used the fact that a polynomial of degree n in $\mathbb{Q}[X]$ is uniquely determined by its values at n+1 distinct points of $\mathbb{Z}^+$. Since distinct weight subspaces are independent, the representation is unique. This proves (b).

We now return to the surjectivity of our map in the proof of (a). Let $\chi \in \mathrm{Hom}_{\mathbb{Z}}(\mathcal{P}^n_i(G)/\mathcal{P}^n_{i+1}(G), Y)$. Since $\mathcal{P}^n_i(G)$ is a $\mathbb{Q}$-subspace of $\mathcal{P}^n(G)$ and Y is a $\mathbb{Q}$-vector space, χ can be extended (in many ways) to $c(\chi)$ in $\mathrm{Hom}_{\mathbb{Z}}(\mathcal{P}^n(G), Y)$. According to (b), we can write $c(\chi) = \Sigma_{1 \leq j \leq n} \chi_j$, where χ_j has $\mathbb{Q}^+$-weight j. According to Proposition 4.3, χ_j is 0 on the subspace $\mathcal{P}^n_i(G)$ when $j < i$. Thus, $c'(\chi) = \Sigma_{i \leq j \leq n} \chi_j$ is also an extension of χ. The same argument shows that the value of $c'(\chi)$ on $\mathcal{P}^n_n(G)$ coincides with that of χ_n. If $i = n$, then $c'(\chi) = \chi_n$ and we are done. If $i < n$, then χ is 0 on $\mathcal{P}^n_n(G)$ so that χ_n is 0 on $\mathcal{P}^n_n(G)$. Proposition 3.3 now shows that $\chi_n = 0$ when $i < n$. This argument can be extended to show that χ_j is 0 for $j > i$. It follows that $c'(\chi) = \chi_i$ is an extension of χ and we have surjectivity. Q.E.D.

Remark 4.5. Assertion (b) of Proposition 4.3 facilitates the computation of invariants which are of $\mathbb{Q}^+$-weight j. For the case of volume, it reduces to the classical formula in the case of a simplex. In a highly nonconstructive sense, any $\mathbb{Q}$-vector space basis for $\mathrm{Hom}_{\mathbb{Z}}(\mathcal{P}^n(G), \mathbb{Q})$ is a complete system of invariants solving the stable G-scissor's congruence problem for $\mathfrak{M}$. Our task is to find a reasonable set of invariants. They do not have to form a $\mathbb{Q}$-vector space basis. They only have to separate the points of $\mathcal{P}^n(G)$. For this reason, we may replace $\mathbb{Q}$ by any convenient $\mathbb{Q}$-vector space Y. The word "reasonable" is to be interpreted to mean that the invariants should be related to the geometry of the polyhedron.

It should be noted that the two sides of the third equation in the proof of Proposition 4.4 make sense for m in k^+ whenever Y is a k-vector space. However, the equality of these two sides is no longer clear. The trouble is that the second canonical decomposition depended on the fact that m is a positive integer. Moreover, equality in the extended situation would force χ_i to have k^+-weight i. In the absence of any additional assumptions on k and χ, this last event is not expected.

5. WEIGHT SPACE DECOMPOSITION.

In view of Proposition 4.4, we can decompose $\mathcal{P}^n(G)$ into a direct sum of weight subspaces with respect to the action of scaling. For $1 \leq i \leq n$, the weight subspace $\mathcal{G}_i(\mathcal{P}^n(G))$ for the weight i is defined to be the $\mathbb{Q}$-subspace formed by all $[P]-[Q]$ such that:

$$[m \circ P] - [m \circ Q] = m^i([P]-[Q]) \text{ in } \mathcal{P}^n(G) \text{ for all m in } \mathbb{Z}^+$$

Proposition 4.4 immediately yields the following assertion:

Proposition 5.1. For any subgroup G between $T_k(\mathfrak{m})$ and $A_k(\mathfrak{m})$, $\mathcal{P}^n(G)$ is the (internal) direct sum of the weight spaces $\mathcal{G}_i(\mathcal{P}^n(G))$, $1 \leq i \leq n$. The weight space $\mathcal{G}_i(\mathcal{P}^n(G))$ is canonically isomorphic to $\mathcal{P}^n_i(G)/\mathcal{P}^n_{i+1}(G)$, $1 \leq i \leq n$.

The preceding proposition shows that $\mathcal{P}^n(G)$ is a graded $\mathbb{Q}$-vector space. The components $\mathcal{G}_i(\mathcal{P}^n(G))$ are stable under scaling by k^+ because k^+ is commutative under multiplication. It follows that the projection operators $\mathcal{G}_i$ are $\mathbb{Q}$-linear and commutes with scaling by k^+. We want to describe $\mathcal{G}_j[A]$ for an arbitrary geometric n-simplex $A = /x_1/\ldots/x_n/$. For this purpose, we define $[A(i)]$ as in Theorem 3.2 (the second canonical decomposition), $1 \leq i \leq n$. When $j > n$, we agree to set $\mathcal{G}_j[A] = 0 = [A(j)]$. The second canonical decomposition can now be written in the form:

$$[m \circ A] = \Sigma_{j>0}\binom{m}{j}[A(j)] = \Sigma_{j>0}\mathcal{G}_j[A]\, m^j \text{ in } \mathcal{P}^n(T_k(\mathfrak{m})),\ m \in \mathbb{Z}^+$$

$\mathcal{G}_i[A]$ is a $\mathbb{Q}$-linear combination of $[A(j)]$, $j > 0$. Specifically, the coefficient of $[A(j)]$ in $\mathcal{G}_i[A]$ is the coefficient of X^i in $\binom{X}{j}$. The precise formula is stated in the following proposition.

Proposition 5.2. Let $\log(1+X)$ be the formal power series $\Sigma_{j>0}(-1)^{j-1}X^j/j$ in $\mathbb{Q}[[X]]$. If $f(X) \in X\mathbb{Q}[[X]]$ is a power series with zero constant term and if A is any geometric n-simplex in $\mathfrak{m}$, then $f(A)$ shall denote the element of $\mathcal{P}^n(G)$ obtained by replacing each X^i, $i > 0$, in $f(X)$ by $[A(i)]$. With this convention, we have:

(a) $\mathcal{G}_i[A] = (\log(1+A)^i)/i!$, $i > 0$;
(b) $\mathcal{G}_i[A] \in \mathcal{P}_i^n(G)$, $i > 0$; and
(c) $\mathcal{G}_i[A] \equiv [A(i)]/i! \bmod \mathcal{P}_{i+1}^n(G)$, $i > 0$.

Proof. (b) and (c) are consequences of (a). Let $\mathbb{Q}[[t]]$ be the formal power series ring in t over $\mathbb{Q}$. Let $\exp(t) = 1 + \Sigma_{i>0}t^i/i!$ and let $\log(1+t)$ be as before. For any g in $\mathbb{Q}[[X, Y]]$ with zero constant term, we can specialize from $\mathbb{Q}[[t]]$ to $\mathbb{Q}[[X, Y]]$ by sending t onto g. With this understanding, we have the following formal identities:

$$\exp(g+h) = \exp(g)\cdot\exp(h) \text{ and } \exp(\log(1+g)) = 1+g.$$

Consider the formal power series $\exp(X\cdot\log(1+Y))$. We have:

$$\exp(X\cdot\log(1+Y)) = 1 + \Sigma_{i>0}\, g_i(Y)X^i = 1 + \Sigma_{i>0}h_i(X)Y^i.$$

It is clear that $g_i(Y) = (\log(1+Y)^i)/i! \in \mathbb{Q}[[Y]]$ has lowest degree term $Y^i/i!$. It follows that $h_i(X) \in \mathbb{Q}[X]$ is a polynomial in X with highest degree term $X^i/i!$. Consequently, $\exp(X\cdot\log(1+Y))$ lies in $\mathbb{Q}[X][[Y]]$ and can be specialized to $\exp(m\cdot\log(1+Y)) = (1+Y)^m = 1 + \Sigma_{i>0}\binom{m}{i}Y^m$ for each m in $\mathbb{Z}^+$. It follows that $h_i(X)$ specializes to $\binom{m}{i}$ for each m in $\mathbb{Z}^+$. Since $h_i(X)$ is a polynomial, we conclude that $h_i(X) = \binom{X}{i}$. The problem of relating the coefficients of g_i and h_j involves a system of linear equations over $\mathbb{Q}$. A comparison of the forms shows that this is the same problem as that involved in relating the coefficients of $[A(j)]$ in $\mathcal{G}_i[A]$ with that of X^i in $\binom{X}{j}$. Assertion (a) follows without difficulty. Q.E.D.

Proposition 5.3. Let P be a polyhedron in $\mathfrak{m}$ and let G be any subgroup between $T_k(\mathfrak{m})$ and $A_k(\mathfrak{m})$. Then $[P] \in \mathcal{P}_{i+1}^n(G)$ if and only if:

$$\chi[P] = 0 \text{ holds for all } \chi \in \mathrm{Hom}_{\mathbb{Z}}(\mathcal{P}^n(G), \mathbb{Q}) \text{ with } \mathbb{Q}^+\text{-weight at most } i.$$

Proof. When $i = n$, the assertion follows from duality between $\mathbb{Q}$-vector spaces $\mathcal{P}^n(G)$ and $\mathrm{Hom}_{\mathbb{Z}}(\mathcal{P}^n(G), \mathbb{Q})$. We now downward induct on i. Assume our assertion is valid for $i = j+1 \leq n$.

Suppose $[P] \in \mathcal{P}^n_{j+1}(G)$. Let χ have $\mathbb{Q}^+$-weight s with $s \leq j$. Since $s \leq j+1$, induction shows that χ is zero on $\mathcal{P}^n_{j+2}(G)$. Proposition 3.6 gives:

$$m^s \chi[P] = m^{j+1} \chi[P], \; m \in \mathbb{Z}^+$$

Since $s \leq j$, we have $\chi[P] = 0$ and our condition is necessary.

Conversely, assume our condition is satisfied. The weight space decomposition together with duality of $\mathbb{Q}$-vector spaces imply that $\mathcal{G}_s[P]$ is 0 for $s \leq j$. By Proposition 5.2, $[P] = \Sigma_{s>j} \mathcal{G}_s[P] \in \mathcal{P}^n_{j+1}(G)$. Q.E.D.

Proposition 5.4. Assume that G contains $T_k(\mathfrak{m})$ and fixes a positive definite inner product $<\,,\,>$ on $\mathfrak{m}$. Then:

(a) $\mathcal{P}^n_i(G)$ is generated by $[P]$ with P ranging over the orthogonal basic i-fold cylinders of $\mathfrak{m}$;

(b) for $i < n$, each element of $\mathcal{P}^n_i(G)/\mathcal{P}^n_n(G)$ is represented as $[\coprod_j P_j] = \Sigma_j [P_j]$ where P_j is a suitable orthogonal basic i-fold cylinder; and

(c) if k is Archimedean and P is a polyhedron in $\mathfrak{m}$ with $[P] \in \mathcal{P}^n_i(G)$, then P is G-scissors congruent to $\coprod_j P_j$ where each P_j is a suitable orthogonal basic i-fold cylinder.

Proof. (a) is a consequence of Theorems 2.4, 3.2 and Propositions 5.1, 5.2. Adapting the proof of Proposition 2.6, (b) follows from (a).

For assertion (c), we use (a) and (b) to obtain:

$$[P] + [\coprod_a R_a] = [\coprod_b S_b] \text{ in } \mathcal{P}^n(G)$$

where each R_a is a suitable orthogonal basic n-fold cylinder while each S_b is a suitable orthogonal basic i-fold cylinder. If $P = \phi$ there is nothing to prove. If $P \neq \phi$, then $\Sigma_b \operatorname{vol} S_b > \Sigma_a \operatorname{vol} R_a$. By using Jordan approximation, we can write S_b as a sum $(\coprod_c S_{b,c}) \coprod (\coprod_d T_{b,d})$ where each $S_{b,c}$ is an orthogonal basic i-fold cylinder, each $T_{b,d}$ is an orthogonal basic n-fold cylinder, and $\Sigma_{b,d} \operatorname{vol} T_{b,d} > \Sigma_a \operatorname{vol} R_a$. According to

Theorem 2.5, we may conclude:

$$\Sigma_{b,d}[T_{b,d}] = \Sigma_a [R_a] + [T] \text{ in } \mathcal{P}^n(G)$$

where T is a suitable orthogonal basic n-fold cylinder. As a consequence, we may assume that each R_a is empty to begin with. Since k is assumed to be Archimedean, (c) now follows from Theorem 3.5. Q.E.D.

The multiplicative group $k^\times$ is the direct product of k^+ and $< \pm 1 >$. We identified k^+ with homotheties on $\mathfrak{m}$ and obtained the weight space decomposition of $\mathcal{P}^n(G)$. We will identify -1 with the central symmetric $-I_n$ of $\mathfrak{m}$. It plays an important role.

Proposition 5.5. Let G be any subgroup between $T_k(\mathfrak{m})$ and $A_k(\mathfrak{m})$. The central symmetry $-I_n$ induces multiplication by $(-1)^{n-i}$ on $\mathcal{G}_i(\mathcal{P}^n(G))$. If $-I_n$ lies in G, then $\mathcal{G}_i(\mathcal{P}^n(G)) = 0$ holds for $n-i \equiv 1 \bmod 2$.

Proof. We have $/x_1/\dots/x_n/ = (x_1+\dots+x_n)+/-x_n/\dots/-x_1/$ for any geometric n-simplex $/x_1/\dots/x_n/$. According to (c) of Proposition 2.1

$$[/x_1/\dots/x_n/] + (-1)^n[/x_n/\dots/x_1/] \in \mathcal{P}^n_2(T_k(\mathfrak{m}))$$

Applying this result to each factor of a basic i-fold cylinder and expand by distributive law, our assertion follows because G contains $T_k(\mathfrak{m})$. Q.E.D.

Remark 5.6. Proposition 5.5 was proved by Hadwiger [31; p. 62] through a combination of a geometric argument with the use of functionals. The present proof is more direct and uses only elementary combinatorics and geometry.

Proposition 5.7. Let n = 2 and assume G contains $T_k(\mathfrak{m}) < \pm I_2 >$ and fixes a positive definite inner product $<\,,\,>$ on $\mathfrak{m}$. Then $\mathcal{P}^2(G) \cong k$ through the area map.

Proof. Use Proposition 5.5 and Theorem 2.5. It is also easy to see this geometrically. Namely, we rotate the "upper half" of a triangle through 180° to make a parallelogram and apply Theorem 2.5. Q.E.D.

Boltianskii [11; Theorem 17, p. 87] has proved a converse of this result.

6. HADIWIGER'S INVARIANTS.

Hadwiger's invariants are suitable functionals on $\mathcal{P}^n(T_k(\mathfrak{M}))$ with values in suitable k-vector spaces (each with value in a 1-dimensional vector space). They are based on the concept of volume. As indicated before, there are two essentially identical procedures to define volume. The first of these requires the choice of an ordered basis of $\mathfrak{M}$ (as well as each of the k-linear subspace of $\mathfrak{M}$). This is the procedure followed by Jessen and Thorup [44] and has the advantage that the values of the functionals all lie in k. The second of these requires the selection of a positive definite inner product on $\mathfrak{M}$ as well as an ordered field extension K of k so that K contains all the square roots of elements of k^+. The values of the functionals will lie in K. This procedure has the advantage that a volume is automatically defined on lower dimensional polyhedra. It also has the added advantage that the transition from the affine to the Euclidean case does not require a serious change in notations. We will follow the second procedure.

For the rest of this section, K will denote a fixed ordered field extension of k so that K contains the square root of every element of k^+. $< \, , \, >$ will denote a fixed positive definite inner product on $\mathfrak{M}$. Each k-linear subspace of $\mathfrak{M}$ will receive the induced inner product. Each i-dimensional polyhedron in $\mathfrak{M}$ will have an i-dimensional (unoriented or absolute) volume as in section 2. If k is square root closed, then K can and will be taken to be k.

A <u>ray</u> r in $\mathfrak{M}$ is defined to be a subset k^+u of $\mathfrak{M}$ for some nonzero vector u in $\mathfrak{M}$. When k is square root closed, there is a unique unit vector u in each ray and we can (and will) identify these concepts. If r_1 and r_2 are rays, then $< r_1, r_2 > = \{ <x_1, x_2> \mid x_i \in r_i \}$ is either 0 or one of the two subsets k^+, k^- of k. When $< r_1, r_2 > = 0$, r_1 and r_2 are said to be <u>orthogonal</u>. An ordered set of pairwise orthogonal rays $r_1, \ldots, r_j$ in $\mathfrak{M}$ will be called an <u>orthogonal j-frame</u> or simply a <u>j-frame</u>.

For each j-frame $r_1, \ldots, r_j$, $0 \leq j \leq n-1$, in $\mathfrak{M}$, the <u>Hadwiger invariant</u> $\Omega^n(r_1, \ldots, r_j)$ is defined in terms of functions $\chi^n(r_1, \ldots, r_j)$ to be described. When k is square root closed, we replace each r_i by the unit vector u_i on r_i.

When j = 0, $\Omega^n = \chi^n$ is defined to be the (absolute) n-dimensional volume on geometric n-simplices and is extended to $\mathcal{P}^n(T_k(\mathfrak{m}))$ through additivity with respect to simple subdivision.

For j > 0, let A denote an arbitrary geometric n-simplex. The function $\chi^n(r_1, \ldots, r_j)$ is defined to be 0 on A unless there is a (necessarily unique) sequence of codimensional i faces $A^{(i)}$:

$$A = A^{(0)} \supset A^{(1)} \supset \ldots \supset A^{(j)}$$

such that each r_i is the exterior normal ray to $A^{(i)}$ in $Asp(A^{(i-1)})$. In that exceptional case, define:

$$\chi^n(r_1, \ldots, r_j)(A) = \Omega^{n-j}(A^{(j)}) = \mathrm{vol}_{n-j}(A^{(j)}) \in K^+.$$

The sequence of faces described above will be called a <u>codimensional 1 face sequence of length j</u>. For j > 0, there are (n+1)...(n+2-j) such sequences for each geometric n-simplex A. It is immediate that the function $\chi^n(r_1, \ldots, r_j)$ is $T_k(\mathfrak{m})$-invariant and has k^+-weight n-j > 0. When j > 0, it is not additive with respect to simple subdivision.

When j > 0, we define $\Omega^n(r_1, \ldots, r_j)$ through the functional equation:

$$\Omega^n(r_1, \ldots, r_j) = \Sigma\, \epsilon_1 \ldots \epsilon_j \chi^n(\epsilon_1 r_1, \ldots, \epsilon_j r_j),\ \epsilon_i = \pm 1,\ 1 \leq i \leq j,$$

where the summation extends over all 2^j possibilities. $\Omega^n(r_1, \ldots, r_j)$ is still $T_k(\mathfrak{m})$-invariant and has k^+-weight n-j > 0. We will now check the additivity of $\Omega^n(r_1, \ldots, r_j)$ with respect to simple subdivision. Once this is accomplished, $\Omega^n(r_1, \ldots, r_j)$ then extends uniquely to an element of $\mathrm{Hom}_{\mathbb{Z}}(\mathcal{P}^n(T_k(\mathfrak{m})), K)$. The checking will be done by induction on j. It is trivial when j = 0. Let B and C be the result from A through the simple subdivision by the cutting hyperplane H. There are several possibilities:

Case 1. H is orthogonal to r_1. r_1 is therefore not orthogonal to any of the codimensional 1 faces of A and we have $\Omega^n(r_1, \ldots, r_j)(A) = 0$. B and C clearly have the common codimensional 1 face $F = A \cap H$. The relevant exterior normal rays are r_1 and $-r_1$. Exchanging B and C when necessary, we obtain:

$$\Omega^n(r_1, \ldots, r_j)(B) + \Omega^n(r_1, \ldots, r_j)(C)$$
$$= \Omega^{n-1}(r_2, \ldots, r_j)(F) - \Omega^{n-1}(r_2, \ldots, r_j)(F) = 0.$$

A comparison of these vanishing results yields the desired additivity.

Case 2. H is not orthogonal to r_1. We may assume r_1 to be orthogonal to the codimensional 1 face F of A. Otherwise, all the relevant terms are 0 and additivity amounts to $0 + 0 = 0$. There are now two possibilities:

Case 2a. F is properly subdivided by H into $F \cap B$ and $F \cap C$. Let $\epsilon_1 r_1$ be the exterior normal ray to F, $\epsilon_1 = \pm 1$. We then have:

$$\Omega^n(r_1, \ldots, r_j)(A) = \epsilon_1 \Omega^{n-1}(r_2, \ldots r_j)(F);$$
$$\Omega^n(r_1, \ldots, r_j)(B) = \epsilon_1 \Omega^{n-1}(r_2, \ldots, r_j)(F \cap B); \text{ and}$$
$$\Omega^n(r_1, \ldots, r_j)(C) = \epsilon_1 \Omega^{n-1}(r_2, \ldots, r_j)(F \cap C).$$

The desired additivity follows by induction.

Case 2b. F is not properly subdivided by H. In this case, F is a codimensional 1 face of exactly one of B and C, say of B. It follows that the third equation in case 2a is just $0 = 0$ and the second equation in case 2a has $B = F \cap B$. The desired additivity is now clear.

With a little care on codimensional 1 face sequences, $\Omega^n(r_1, \ldots, r_j)$ can be defined directly on polyhedra of dimension n in $\mathfrak{M}$. For geometric n-simplices, the function $\chi^n(r_1, \ldots, r_j)$ can be bypassed. Namely, if A is any geometric n-simplex, then $\Omega^n(r_1, \ldots, r_j)(A) = 0$ unless there is a (necessarily unique) sequence of codimensional i face $A^{(i)}$ with exterior normal ray $\epsilon_i r_i$ in $\mathrm{Asp}(A^{(i-1)})$, $\epsilon_i = \pm 1$. In that case, $\Omega^n(r_1, \ldots, r_j)(A)$ is just $\epsilon_1 \cdot \ldots \cdot \epsilon_j \cdot \mathrm{vol}(A^{(j)})$. The intermediate function $\chi^n(r_1, \ldots, r_j)$ clearly exhibits the alternating character of $\Omega^n(r_1, \ldots, r_j)$:

$$\Omega^n(\eta_1 r_1, \ldots, \eta_j r_j) = \eta_1 \ldots \eta_j \Omega^n(r_1, \ldots, r_j), \ \eta_i = \pm 1.$$

For a fixed j, $0 < j < n$, there are other nontrivial relations among the Hadwiger invariants. This is the problem of syzygy. The intermediate functions $\chi^n(r_1, \ldots, r_j)$ can be used in defining some of the maps.

3 Translational scissors congruence

Let $\mathfrak{m}$ be the affine n-space over an ordered field k. Let P # Q be any basic 2-fold cylinder in $\mathfrak{m}$ and let $\lambda \in k^+$. The principal result in the present chapter is the assertion:

$$[(\lambda \circ P) \# Q] \equiv [P \# (\lambda \circ Q)] \bmod \mathcal{P}_3^n(T_k(\mathfrak{m}))$$

The assertion is vacuously valid when $n < 2$. When $n = 2$, it follows from Theorem 2.2.5. For $n = 3$, the assertion is due to Hadwiger [33] with $k = \mathbb{R}$. Jessen then proved it for $n = 3$ and 4 [41]. According to the manuscript [44], the general assertion is due to A. Thorup in 1972-73. Without being aware of the content of the aforementioned manuscript, we found a slightly different proof in 1977 [60] for the case where k is Archimedean. With a minor modification, our proof also works for any ordered field k. The proof presented in this chapter follows the line of reasoning presented in our preliminary manuscript [60]. On the surface, the present proof is more algebraic and combinatorial than the proof of Thorup. In reality, the algebra and the combinatorics are merely book-keeping devices for the geometry of translational scissors congruence (the affine case). The more geometrically oriented readers should consult the references mentioned for motivations as well as for a different proof. Throughout the present chapter, the group G of motions will always be the group of all translations. We will occasionally consider k-subspaces $\mathfrak{n}$ of dimension m in $\mathfrak{m}$. When there is no chance of confusion, we abbreviate $\mathcal{P}_i^m(T_k(\mathfrak{n}))$ and $\mathcal{G}_i(\mathcal{P}^m(T_k(\mathfrak{n})))$ to $\mathcal{P}_i^m(\mathfrak{n})$ and $\mathcal{G}_i^m(\mathfrak{n})$ or even to $\mathcal{P}_i^m$ and $\mathcal{G}_i^m$. The set of all ordered basis of $\mathfrak{n}$ over k will be denoted by $GL_k(\mathfrak{n})$. We will also view $GL_k(\mathfrak{n})$ as the group of all k-linear automorphisms of $\mathfrak{n}$ so that $GL_k(\mathfrak{n})$ is a group acting regularly via left and right multiplication on the set $GL_k(\mathfrak{n})$ of all ordered basis of $\mathfrak{n}$.

1. BEGINNING OF THE PROOF.

Let $(x_1, \ldots, x_n) \in GL_k(\mathfrak{m})$, $n \geq 2$. For $\lambda \in k^+$ and $1 \leq j \leq n-1$, we define $D_{\lambda,j}(x_1, \ldots, x_n) \in \mathcal{P}_2^n(\mathfrak{m})/\mathcal{P}_3^n(\mathfrak{m}) \cong \mathcal{G}_2^n$ to be the coset of:

$$[(\lambda o/x_1/\ldots/x_j/)\#/x_{j+1}/\ldots/x_n/] - [/x_1/\ldots/x_j/\#(\lambda o/x_{j+1}/\ldots/x_n/)]$$

The main result now reads:

Theorem 1.1. $D_{\lambda,j}(x_1, \ldots, x_n)$ is always 0.

As a first step towards the proof of Theorem 1.1, we will show:

Lemma 1.2. $\Sigma_{1 \leq j \leq n-1} D_{\lambda,j}(x_1, \ldots, x_n)$ is always 0.

We observe that Lemma 1.2 is just Theorem 1.1 when n = 2 and follows from Theorem 2.2.5 in this case. We may assume n > 2. In general, $D_{\lambda,j}(x_1, \ldots, x_n)$ depends only on λ and the geometric simplices $/x_1/\ldots/x_j/$ and $/x_{j+1}/\ldots/x_n/$ (up to translations). It does not depend on the particular representations of these simplices. In this respect, it is not surprising that Theorem 1.1 follows from Lemma 1.2. However, the proof is rather complicated.

Let $\mu = \lambda^{-1} \in k^+$. It is immediate that:

$$D_{\lambda,j}(x_1, \ldots, x_n) = -D_{\mu,j}(\lambda x_1, \ldots, \lambda x_n)$$

We may therefore assume $\lambda \geq 1$ in the verification of Lemma 1.2. It is also immediate that:

$$D_{1,j}(x_1, \ldots, x_n) = 0 \text{ and } D_{\lambda+m,j}(x_1, \ldots, x_n) = D_{\lambda,j}(x_1, \ldots, x_n),\ m \in \mathbb{Z}^+$$

In case k is Archimedean, we may assume $1 < \lambda < 2$. Since k is only taken to be ordered, we may not assume $1 < \lambda < 2$. However, we can assume $\lambda > 2$.

Let x, y, $z_1, \ldots, z_m$ be a k-basis of $\mathfrak{m}$, m = n-2 > 0. The basic 2-fold cylinder $/x/\#/y/z_1/\ldots/z_m/$ can be viewed as a portion of an affine fibration over the xy-plane. Over the point $\alpha x+\beta y$, $\alpha, \beta \geq 0$, the fibre is just the geometric m-simplex $(\alpha x+\beta y) + (\beta o/z_1/\ldots/z_m/)$ (it degenerates to a point when $\beta = 0$, see also section 2 of Chapter 2). Each of the two polyhedra appearing in $D_{\lambda,1}(x, y, z_1, \ldots, z_m) = D_{\lambda,1}(-x, y, z_1, \ldots, z_m)$ can be taken to be portion of the fibration sitting over the rectangles as indicated in the diagram (a). If $pr_{x,y}$ denotes the projection from $Lsp(x, y, z_1, \ldots, z_m)$ onto $Lsp(x, y)$ with kernel $Lsp(z_1, \ldots, z_m)$ and if $\lambda > 2$, then $D_{\lambda,1}(x, y, z_1, \ldots, z_m)$ is the sum of the following two differences: (diag. (a) shows the bases.)

$$[pr_{x,y}^{-1}((y+(\lambda-1)x)+/x-y/y/)] - [pr_{x,y}^{-1}(\lambda y+/x-y/y/)] \tag{1.3}$$

$$[pr_{x,y}^{-1}(x+(\lambda-1)\,o\,/y/x-y/)] - [pr_{x,y}^{-1}(y+(\lambda-1)\,o\,/y/x-y/)] \tag{1.4}$$

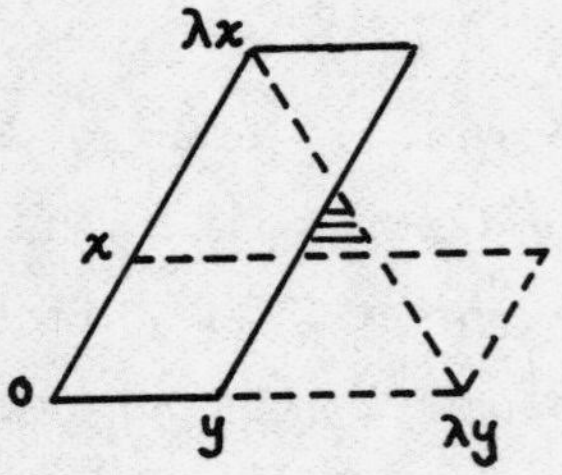

(a)

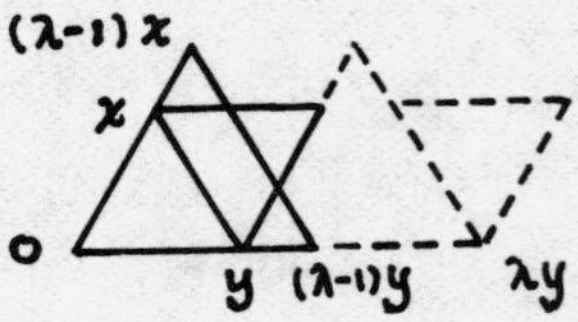

(b)

Note that the portion over the shaded triangle is first added on then subtracted off in achieving (1.4). This would not be necessary if $1 < \lambda < 2$. When k is Archimedean, we may assume $1 < \lambda < 2$.

We next note: the fibre $(\alpha x+\beta y) + \beta\, o/z_1/\ldots/z_m/$ over $\alpha x+\beta y$, $\alpha, \beta \geq 0$, is translated by $(\gamma-\alpha)x$ exactly onto the fibre $(\gamma x+\beta y) + \beta\, o/z_1/\ldots/z_m/$ over $\gamma x+\beta y$, $\gamma, \beta \geq 0$. It follows that (1.3) and (1.4) are respectively equal to (1.5) and (1.6): (diagram (b) above shows the bases.)

$$[pr_{x,y}^{-1}(y+/x-y/y/)] - [pr_{x,y}^{-1}(\lambda y+/x-y/y/)] \tag{1.5}$$

$$[pr_{x,y}^{-1}((\lambda-1)\,o/y/x-y/)] - [pr_{x,y}^{-1}(y+(\lambda-1)\,o/y/x-y/)] \tag{1.6}$$

The triangles $y+/x-y/y/$, $/y/x-y/$ make up the parallelogram $/x/\#/y/$. Similarly, $\lambda y+/x-y/y/$, $(\lambda-1)y+/y/x-y/$ make up $(\lambda-1)y+(/x/\#/y/)$ (diag. (b)).

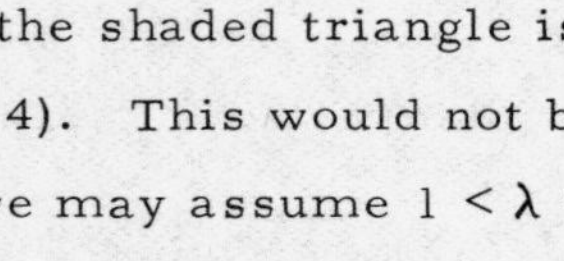

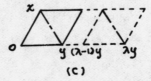

(c)

Consider the difference: (see diagram (c) on preceding page.)

$$[pr_{x,y}^{-1}(/x/\#/y/)] - [pr_{x,y}^{-1}((\lambda-1)y+(/x/\#/y/))] \qquad (1.7)$$

Since the base is a product and the total space is also a product and is compatible with the product in the base, the expression (1.7) is equal to:

$$[/x/\#P] \qquad (1.8)$$

where P is the convex closure in $Lsp(y, z_1, \ldots, z_m)$ of the following set:

$$pr_y^{-1}((\lambda-1)y+/y/) - \{(\lambda-1)y + pr_y^{-1}(/y/)\} \qquad (1.9)$$

and pr_y is the projection of $Lsp(y, z_1, \ldots, z_m)$ onto $Lsp(y) = ky$ with kernel $Lsp(z_1, \ldots, z_m)$ (see diagram (d)). Since we are dealing with translational scissors congruence up to stability, the set $(\lambda-1)y + pr_y^{-1}(/y/)$ can be translated to $(\lambda-1)(y+z_1+\ldots+z_m)+pr_y^{-1}(/y/)$ (see diagram (e)). We now apply the first canonical decomposition to $\lambda \circ /y/z_1/\ldots/z_m/$ with $\lambda = (\lambda-1) + 1$. It

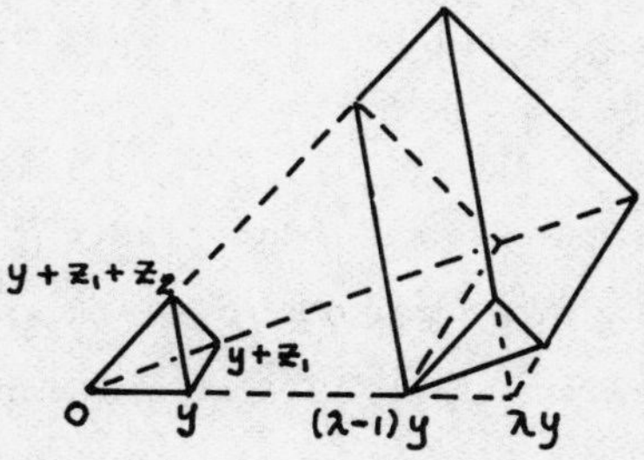

(d)

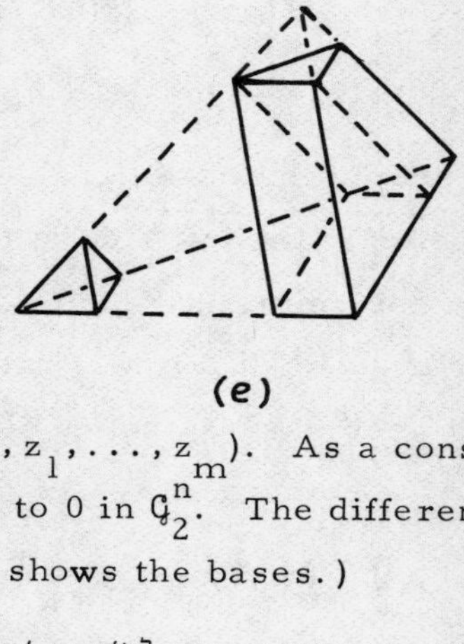
(e)

follows that P is a 2-fold cylinder in $Lsp(y, z_1, \ldots, z_m)$. As a consequence, the expression in (1.8) lies in $\mathcal{P}_3^n$ and leads to 0 in $\mathcal{G}_2^n$. The difference in (1.5) can now be replaced by: (diagram (f) shows the bases.)

$$-[pr_{x,y}^{-1}(/y/x-y/)] + [pr_{x,y}^{-1}((\lambda-1)y+/y/x-y/)] \qquad (1.10)$$

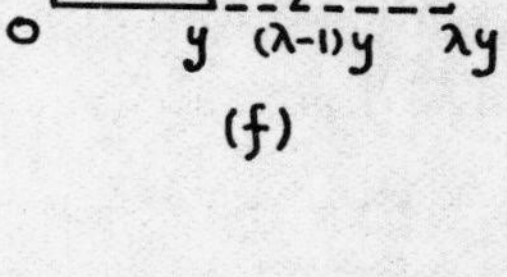

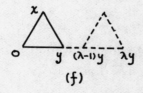

(f)

We observe that (1.6) and (1.10) both involve the calculation of the following expression:

$$[pr^{-1}_{x,y}(hy + t\circ/y/x-y/)] - [pr^{-1}_{x,y}(t\circ/y/x-y/)] \qquad (1.11)$$

The two cases correspond to (h, t) = (1, λ-1) and (λ-1, 1) respectively. It is evident that:

$$h(y+z_1+\ldots+z_m)+pr^{-1}_{x,y}(t\circ/y/x-y/) \subset pr^{-1}_{x,y}(hy+t\circ/y/x-y/) \qquad (1.12)$$

Let C denote the convex closure of the difference between the two sides of (1.12). It is evident that [C] is equal to the expression in (1.11).

We now examine [C]. Let $\alpha x+\beta y \in t\circ/y/x-y/$. The fiber over the point $\alpha x + (\beta+h)y$ is equal to $(\beta+h)\circ/z_1/\ldots/z_m/$. The first canonical decomposition of the fiber yields:

$$\begin{aligned}(h+\beta)\circ/z_1/\ldots/z_m/ &= h\circ/z_1/\ldots/z_m/ \\ &\amalg\, hz_1+\{(\beta\circ/z_1/)\#(h\circ/z_2/\ldots/z_m/)\} \amalg \ldots \qquad (1.13)\\ &\amalg\, h(z_1+\ldots+z_m)+\{\beta\circ/z_1/\ldots/z_m/\}\end{aligned}$$

The last term in the total fiber decomposition (1.13) is precisely the part corresponding to the left side of (1.12) and sits over $\alpha x+(\beta+h)y$ in the base. Moreover, the relevant translations appearing in (1.13) are all along the fiber coordinates. It follows that [C] is the sum of the classes of m = n-2 polyhedra; each of these polyhedra is a fibration over $t\circ/y/x-y/$ so that the fiber over the point $\alpha x+\beta y$ has the form:

$$(\alpha x+\beta y)+\{(\beta\circ/z_1/\ldots/z_j/)\#(h\circ/z_{j+1}/\ldots/z_m/)\},\ 0\le j\le m-1. \quad (1.14)$$

According to section 2 of Chapter 2, [C] is equal to:

$$\Sigma_{0\le j\le m-1}[(t\circ/-x/y/z_1/\ldots/z_j/)\ \#\ (h\circ/z_{j+1}/\ldots/z_m/)] \qquad (1.15)$$

We note that /y/x-y/ = x + /-x/y/ and that translation by -x has no effect on the description of the fibration because Lsp(x) = kx is a factor in the total fibration. This allows us to use /-x/y/ in place of /y/x-y/ so that the description in section 2 of Chapter 2 can be used without encountering problems.

Application of (1.15) to $(h, t) = (1, \lambda-1)$, $(\lambda-1, 1)$ and subtract correctly yield:

$$D_{\lambda,1}(-x, y, z_1, \ldots, z_m) = -\Sigma_{2 \leq j \leq n-1} D_{\lambda-1,j}(-x, y, z_1, \ldots, z_m) \quad (1.16)$$

Since $D_{\lambda-1,j} = D_{\lambda,j}$ holds, (1.16) yields Lemma 1.2.

We could have started with $D_{\lambda,p}$ for any p between 1 and n-1. Except when $p = n-1$, the results would be more complicated in appearance. In fact, they are equivalent to the case $p = 1$.

2. CONCLUSION OF THE PROOF.

The derivation of Theorem 1.1 from Lemma 1.2 is motivated by the proof of Hadwiger [33] for $n = 3$ and the proof of Jessen [41] for $n = 3, 4$. The formalism emerged after we have verified Theorem 1.1 for $n \leq 7$.

Let Y be a $\mathbb{Q}$-vector space. Y may stand for $\mathbb{Q}$, k or even $\mathcal{G}_2^n$. We recall Proposition 2.4.1 and identify $\mathrm{Hom}_{\mathbb{Z}}(\mathcal{P}^n(\mathfrak{m}), Y)$ with a subset of $\mathrm{Map}(GL_k(\mathfrak{m}), Y)$. From this point of view, $D_{\lambda,j}$ lies in $\mathrm{Map}(GL_k(\mathfrak{m}), \mathcal{G}_2^n)$. If $x = (x_1, \ldots, x_n)$ and $\sigma \in GL_k(\mathfrak{m})$, then $\sigma \cdot x = (\sigma x_1, \ldots, \sigma x_n) \in GL_k(\mathfrak{m})$. For any $f \in \mathrm{Map}(GL_k(\mathfrak{m}), Y)$, let $\sigma f \in \mathrm{Map}(GL_k(\mathfrak{m}), Y)$ be defined by:

$$(\sigma f)(x) = f(\sigma^{-1} \cdot x)$$

It is easy to see that $\mathrm{Map}(GL_k(\mathfrak{m}), Y)$ becomes a left $\mathbb{Q}GL_k(\mathfrak{m})$-module with $\mathrm{Hom}_{\mathbb{Z}}(\mathcal{P}^n(\mathfrak{m}), Y)$ as a $\mathbb{Q}GL_k(\mathfrak{m})$-submodule. In a similar way, $GL_k(\mathfrak{m})$ also acts on itself through the composition of inversion followed by right multiplication. This will in general not stabilize $\mathrm{Hom}_{\mathbb{Z}}(\mathcal{P}^n(\mathfrak{m}), Y)$. However, certain special cases will be of interest.

Let $g \in \mathrm{Map}(GL_k(\mathfrak{m}), Y)$. For $1 \leq i \leq n$, let $\mathfrak{a}_i g \in \mathrm{Map}(GL_k(\mathfrak{m}), Y)$ be defined by the following formula:

$$(\mathfrak{a}_i g)(x_1, \ldots, x_n) = (g(x_1, \ldots, x_n) - g(x_1, \ldots, -x_i, \ldots, x_n))/2$$

It is immediate clear that:

$$\mathfrak{a}_i \circ \mathfrak{a}_j = \mathfrak{a}_j \circ \mathfrak{a}_i \quad \text{and} \quad \mathfrak{a}_i \circ \mathfrak{a}_i = \mathfrak{a}_i$$

Indeed, $\mathfrak{a}_i$ is a projection operator of the $\mathbb{Q}GL_k(\mathfrak{m})$-module $\mathrm{Map}(GL_k(\mathfrak{m}), Y)$ arising from right multiplication by $\mathrm{diag}(1, \ldots, -1, \ldots, 1)$ where -1 is in the i-th position. We note also that the shuffling process is an integral linear combination of the right multiplication action by permutation matrices. The collection of diagonal matrices in $GL(n;\mathbb{Z})$ and the collection of permutation matrices in $GL(n;\mathbb{Z})$ together generate the monomial subgroup of $GL(n;\mathbb{Z})$. It is isomorphic to the Weyl group of type C_n (or B_n). The rational group algebra of the monomial group acts as endomorphisms of the $\mathbb{Q}GL_k(\mathfrak{M})$-module $\mathrm{Map}(GL_k(\mathfrak{M}), Y)$ and our projection operator $\mathfrak{A}_i$ is just an idempotent in this rational group algebra of the monomial group. An essential part of the translational scissors congruence problem is contained in the analysis of the action of these projection operators $\mathfrak{A}_i$ on the subset $\mathrm{Hom}_{\mathbb{Z}}(\mathcal{P}^n(\mathfrak{M}), Y)$.

Lemma 2.1. Let $f \in \mathrm{Hom}_{\mathbb{Z}}(\mathcal{P}^n(\mathfrak{M}), Y)$. Then $\mathfrak{A}_1 \cdots \mathfrak{A}_n f = 0$.

Proof. We induct on n. If $n = 1$, then $f(x_1) = f([/x_1/]) = f(-x_1)$ so that $\mathfrak{A}_1 f = 0$. Assume the assertion is valid for all $m < n$ and let $n > 1$. We note that $2^n \mathfrak{A}_1 \ldots \mathfrak{A}_n f(x_1, \ldots, x_n)$ is a sum of 2^n terms of the form:

$$\epsilon_1 \ldots \epsilon_n f(\epsilon_1 x_1, \ldots, \epsilon_n x_n), \quad \epsilon_i = \pm 1.$$

These terms can be collected into 2^{n-1} groups of the form:

$$f(y_1, \ldots, y_n) + (-1)^n f(-y_1, \ldots, -y_n), \quad y_i = \pm x_i.$$

According to (c) of Proposition 2.2.1, we have:

$$[/y_1/\ldots/y_n/] + (-1)^n [/y_n/\ldots/y_1/]$$
$$= \Sigma_{1 \leq i \leq n-1} (-1)^{i-1} [/y_i/\ldots/y_1/\#/y_{i+1}/\ldots/y_n/] \text{ in } \mathcal{P}^n(\mathfrak{M}).$$

Since $/y_n/\ldots/y_1/ = (y_1 + \ldots + y_n) + /-y_1/\ldots/-y_n/$, it follows that $\mathfrak{A}_1 \ldots \mathfrak{A}_n f(x_1, \ldots, x_n)$ is a $\mathbb{Q}$-linear combinations of terms of the form:

$$f(/y_i/\ldots/y_1/\#/y_{i+1}/\ldots/y_n/), \quad y_j = \pm x_j, \quad 1 \leq i \leq n-1.$$

In this expression, we use the assumption $f \in \mathrm{Hom}_{\mathbb{Z}}(\mathcal{P}^n(\mathfrak{M}), Y)$ so that f can be evaluated on any polyhedron in $\mathfrak{M}$. For a fixed choice of $y_{i+1}, \ldots, y_n$, let $\mathfrak{N} = \mathrm{Lsp}(y_1, \ldots, y_i)$. Define $g \in \mathrm{Map}(GL_k(\mathfrak{N}), Y)$ by:

$$g(w_1, \ldots, w_i) = f(/w_i/\ldots/w_1/\#/y_{i+1}/\ldots/y_n/).$$

From Proposition 2.4.1, it is evident that $g \in \mathrm{Hom}_{\mathbb{Z}}(\mathcal{P}^i(\mathfrak{N}), Y)$. Collecting all the terms with $w_j = \pm x_j$, it is then clear that $\mathfrak{a}_1 \dots \mathfrak{a}_n f(x_1, \dots, x_n)$ is a $\mathbb{Q}$-linear combination of terms of the form:

$$\mathfrak{a}_1 \dots \mathfrak{a}_i g(w_1, \dots, w_i), \quad 1 \leq i \leq n-1$$

By induction, each such term is 0 so that $\mathfrak{a}_1 \dots \mathfrak{a}_n f = 0$. Q.E.D.

Lemma 2.2. Let $f \in \mathrm{Hom}_{\mathbb{Z}}(\mathcal{P}^n(\mathfrak{m}), Y)$ and assume $\mathfrak{a}_n f = 0$. Then:

(a) $\mathfrak{a}_1 f = 0$;

(b) $f(2^s x_1, \dots, 2^s x_n) = 2^{ns} f(x_1, \dots, x_n)$, $s \in \mathbb{Z}$; and

(c) f has $\mathbb{Q}^+$-weight n.

Proof. (a) follows from $/x_1/\dots/x_n/ = (x_1+\dots+x_n) + /-x_n/\dots/-x_1/$. (c) follows from (b) and Proposition 2.4.4.

In proving (b), we may take $s = 1$. Since $/2x_1/ = /x_1/ \coprod (x_1 + /x_1/)$, (b) holds for $n = 1$. Assume $n > 1$, we assert:

$$[/x_1/\dots/x_n/] + [/x_1/\dots/x_{n-1}/-x_n/]$$
$$= [/x_1/\dots/x_{n-2}/x_{n-1}-x_n/2x_n/]$$
$$= [/x_1/\dots/x_{n-2}/x_{n-1}+x_n/-2x_n/] \quad \text{in } \mathcal{P}^n(\mathfrak{m})$$

In fact, up to translations, the first two geometric n-simplices result from either of the latter two through a simple bisection. The assumptions on f translate to the assertion:

$f[A] = 2f[B] = 2f[C]$, where $A = B \coprod C$ is a simple bisection of geometric n-simplices.

A suitable sequence of n simple bisections of the n-simplex $2 \circ A$ will yield A. (b) follows from the preceding assertion by iteration. Q.E.D.

Lemma 2.3. Let $f \in \mathrm{Hom}_{\mathbb{Z}}(\mathcal{P}^n(\mathfrak{m}), Y)$ and assume $\mathfrak{a}_s \dots \mathfrak{a}_n f = 0$ for some s. Let $z_s, \dots, z_n \in \mathfrak{m}$ be k-linearly independent and let $\mathfrak{n}$ be any k-subspace of $\mathfrak{m}$ complementary to $\mathrm{Lsp}(z_s, \dots, z_n)$. Define $g \in \mathrm{Map}(GL_k(\mathfrak{n}), Y)$ by:

$$g(w_1, \dots, w_{s-1}) = \mathfrak{a}_{s+1} \dots \mathfrak{a}_n f(w_1, \dots, w_{s-1}, z_s, \dots, z_n)$$

Then $g \in \mathrm{Hom}_{\mathbb{Z}}(\mathcal{P}^{s-1}(\mathfrak{n}), Y)$.

Proof. According to Proposition 2.4.1, there are two conditions to be verified. We first verify (b), the additivity with respect to simple subdivision of $w = (w_1, \ldots, w_{s-1}) \in GL_k(\mathfrak{h})$. If we subdivide on the i-th entry w_i with $i < s-1$, then additivity of g follows from that of f. The point is the subdivision on w_i involves at most the (i-1)-th, i-th and (i+1)-th entries of w while the operators $\mathfrak{a}_{s+1}, \ldots, \mathfrak{a}_n$ do not involve any of the entries of w. It remains for us to verify additivity when we subdivide on w_{s-1}. Let α, β lie in k^+ with $\alpha + \beta = 1$. We note that $t = \beta/\alpha = (1-\alpha)/\alpha$ so that $\alpha = (1+t)^{-1}$. Thus, t ranges over k^+ as α ranges over k^+ with $0 < \alpha < 1$; clearly $\pm 2t$ then ranges over all of $k^\times$. Simple subdivision at αw_{s-1} then yields:

$$
\begin{aligned}
&\mathfrak{a}_{s+1} \ldots \mathfrak{a}_n f(\ldots, w_{s-2}, w_{s-1}, z_s, \ldots) \\
&\quad = \mathfrak{a}_{s+1} \ldots \mathfrak{a}_n f(\ldots, w_{s-2} + \alpha_{s-1}, \beta w_{s-1}, z_s, \ldots) \\
&\quad + \mathfrak{a}_{s+1} \ldots \mathfrak{a}_n f(\ldots, w_{s-2}, \alpha w_{s-1}, \beta w_{s-1} + z_s, \ldots)
\end{aligned}
\tag{2.4}
$$

We have left out the unaffected j-th entries for $j \neq s-2$, $s-1$ and s. We now use the assumption $\mathfrak{a}_s \mathfrak{a}_{s+1} \ldots \mathfrak{a}_n f = 0$ several times. It is immediate that the left side and the first term of the right side of (2.4) are unchanged when z_s is replaced by $-z_s$. As a consequence, the second term on the right side of (2.4) is unchanged if we replace z_s by $-z_s$ or if we replace $\beta w_{s-1} + z_s$ by $-(\beta w_{s-1} + z_s)$; combining these, it is still unchanged if we replace $\beta w_{s-1} + z_s$ by $-\beta w_{s-1} + z_s$. If we now replace w_{s-1} by $\alpha^{-1} w_{s-1}$, set $t = \beta/\alpha$ and replace $tw_{s-1} \pm z_s$ by z_s, we may then conclude that:

$$
\begin{aligned}
&\mathfrak{a}_{s+1} \ldots \mathfrak{a}_n f(\ldots, w_{s-2}, w_{s-1}, z_s, \ldots) \\
&\quad = \mathfrak{a}_{s+1} \ldots \mathfrak{a}_n f(\ldots,, w_{s-2}, w_{s-1}, tw_{s-1} \pm z_s, \ldots), \ t \in k
\end{aligned}
\tag{2.5}
$$

Feeding (2.5) back to the second term on the right side of (2.4), we obtain the additivity of g with respect to simple subdivision of $w \in GL_k(\mathfrak{h})$.

For condition (a) of Proposition 2.4.1, we note that the geometric n-simplex $-(w_1 + \ldots + w_{s-1} + z_s) + /w_1/\ldots/w_{s-1}/z_s/\ldots/z_n/$ is the geometric join of the geometric (n-s)-simplex $/z_{s+1}/\ldots/z_n/$ and the geometric (s-1)-simplex $-(w_1 + \ldots + w_{s-1} + z_s) + /w_1/\ldots/w_{s-1}/ = -z_s + /-w_{s-1}/\ldots/-w_1/$. Without using any translations, the geometric (s-1)-simplex

$/-w_{s-1}/\ldots/-w_1/$ has $(s-2)!$ distinct nonhomogeneous representations as $/v_1/\ldots/v_{s-1}/$ where v_1 can be taken to range over a k-basis of $\mathfrak{n}$. For example, $/v_1/\ldots/v_{s-1}/ = \mathrm{ccl}\{0, v_1, \ldots, v_1+\ldots+v_i, \ldots, v_1+\ldots+v_{s-1}\}$. By exchanging v_1 with $v_1+\ldots+v_i$, the associated nonhomogeneous representation will have $v_1+\ldots+v_i$ in place of v_1. Equation (2.5) now implies that g can be unambiguously defined so as to be constant on the $T_k(\mathfrak{n})$ orbit of $/w_1/\ldots/w_s/$. Specifically, (2.5) shows that the geometric (s-1)-simplex $-(w_1+\ldots+w_{s-1}+z_s)+/w_1/\ldots/w_{s-1}/ = -z_s+/-w_{s-1}/\ldots/-w_1/$ can be replaced by its translate under tw_{s-1}, $t \in k$. Since w_{s-1} ranges over a k-basis of $\mathfrak{n}$, the desired assertion follows. We have therefore verified (a) of Proposition 2.4.1 for g. Q.E.D.

We will next single out certain formal properties of $D_{\lambda,j}$, $1 \le j \le n-1$, and $\lambda \in k^+$. For this purpose, let $1 \le i \le n$ and let $\eta = \pm 1$. Let $\mathfrak{F}(i,\eta)$ be the $\mathbb{Q}$-subspace of $\mathrm{Map}(GL_k(\mathfrak{m}), Y)$ formed by all maps f with the following properties:

($\mathfrak{F}_1$) $f(x_1, \ldots, x_n) = f(y_1, \ldots, y_n)$ holds whenever:

(a) $\mathrm{Lsp}/x_1/\ldots/x_i/ = \mathrm{Lsp}/y_1/\ldots/y_i/ = F$, $\mathrm{Lsp}/x_{i+1}/\ldots/x_n/ = \mathrm{Lsp}/y_{i+1}/\ldots/y_n/ = B$; and

(b) $/x_1/\ldots/x_i/$ and $/y_1/\ldots/y_i/$ are congruent under $T_k(F)$ and $/x_{i+1}/\ldots/x_n/$ and $/y_{i+1}/\ldots/y_n/$ are congruent under $T_k(B)$;

($\mathfrak{F}_2$) f is biadditive with respect to simple subdivisions of $(x_1, \ldots, x_i)$ and of $(x_{i+1}, \ldots, x_n)$;

($\mathfrak{F}_3$) if $/x_1(j)/\ldots/x_i(j)/$ are geometric i-simplices in F so that $\Sigma_j[/x_1(j)/\ldots/x_i(j)/] \in \mathcal{P}^i_2(F)$, then for any k-basis $x_{i+1}, \ldots, x_n$ of any subspace B complementary to F, we have:

$$\Sigma_j f(x_1(j), \ldots, x_i(j), x_{i+1}, \ldots, x_n) = 0$$

($\mathfrak{F}_4$) if $i < n$, and $(x_1, \ldots, x_n) \in GL_k(\mathfrak{m})$, then:

$$f(x_1, \ldots, x_i, -x_{i+1}, \ldots, -x_n) = \eta\cdot(-1)^{n-i-1} f(x_1, \ldots, x_n)$$

Lemma 2.4. For $1 \le j \le n-1$, $\lambda \in k^+$, $D_{\lambda,j} \in \mathfrak{F}(j,1)$ with $Y = \mathfrak{G}^n_2$.

Proof. Properties ($\mathfrak{F}_1$), ($\mathfrak{F}_2$) and ($\mathfrak{F}_3$) are direct consequences of definition of $D_{\lambda,j}$ and the distributive law of Minkowski sum with respect to interior disjoint union. According to Proposition 2.5.4, $-I_{n-j}$ acts as multiplication by $(-1)^{n-j-1}$ on $\mathcal{G}_1^{n-j}$. Condition $(\mathfrak{F}_4)$ with $\eta = 1$ is therefore a consequence of the fact that $D_{\lambda,j}$ satisfies the symmetrized condition $(\mathfrak{F}_3)$.

Q.E.D.

Lemma 2.5. Fix a choice of η. The $\mathbb{Q}$-subspaces $\mathfrak{F}(i,\eta)$, $1 \leq i \leq n$ are independent, i.e., if $f_i \in \mathfrak{F}(i,\eta)$, then $\Sigma_i f_i = 0$ in $\mathrm{Map}(GL_k(\mathfrak{m}), Y)$ if and only if each f_i is 0 as a map.

Proof. We proceed by induction on n. When $n = 1$, there is nothing to prove. We assume the assertion has been verified for all $m < n$. Let $n > 1$ and let $f_i \in \mathfrak{F}(i,\eta)$ so that $\Sigma_i f_i = 0$, $1 \leq i \leq n$.

As a first step, we show that $f_n = 0$. There are two cases.

Case 1. $\eta = 1$. From Proposition 2.2.1, we obtain:

$$[/x_1/\ldots/x_i/] \equiv (-1)^i [/-x_1/\ldots/-x_i/] \bmod \mathcal{P}_2^i$$

From $(\mathfrak{F}_3)$ and $(\mathfrak{F}_4)$ for $\mathfrak{F}(i,\eta)$, we obtain:

$$f_i(-x_1,\ldots,-x_n) = (-1)^{n-2} f_i(x_1,\ldots,x_n), \quad i < n$$
$$f_n(-x_1,\ldots,-x_n) = (-1)^{n-1} f_n(x_1,\ldots,x_n)$$

Since $\Sigma_i f_i = 0$ and Y is a $\mathbb{Q}$-vector space, we have $f_n = 0$.

Case 2. $\eta = -1$. Let $m \in \mathbb{Z}^+$. From conditions $(\mathfrak{F}_1)$, $(\mathfrak{F}_2)$ and $(\mathfrak{F}_3)$ on $\mathfrak{F}(i,\eta)$ together with the first canonical decomposition, we have:

$$f_i(mx_1,\ldots,mx_n) = mf_i(x_1,\ldots,x_i,mx_{i+1},\ldots,mx_n) \tag{2.6}$$

It follows that $f_n(mx_1,\ldots,mx_n) = mf_n(x_1,\ldots,x_n)$ is linear in m. We assert that f_j, $1 \leq j \leq n-1$, are polynomials of degree at most $n-j$ without constant or linear terms in m and with coefficients in a finite dimensional $\mathbb{Q}$-subspace of Y depending only on $x_1,\ldots,x_n$. Once this is verified, the assumption $\Sigma_i f_i = 0$ will imply $f_n = 0$. To prove the assertion about f_j, fix $x_1,\ldots,x_j$ and view f_j as a function g_{n-j} on $GL_k(\mathfrak{n})$ where $(x_{j+1},\ldots,x_n)$ ranges over $GL_k(\mathfrak{n})$ and $\mathfrak{n}$ is an arbitrary but fixed subspace of $\mathfrak{m}$ complementary to $\mathrm{Lsp}(x_1,\ldots,x_j)$. Conditions $(\mathfrak{F}_1)$, $(\mathfrak{F}_2)$ and $(\mathfrak{F}_4)$ on $\mathfrak{F}(j,-1)$ now imply the following statements:

$$g_t \in \mathrm{Hom}_{\mathbb{Z}}(\wp^t(\mathfrak{h}), Y), \quad 1 \leq t \leq n-1 \tag{2.7}$$

$$g_t(-y_1, \ldots, -y_t) = (-1)^t g_t(y_1, \ldots, y_t), \quad 1 \leq t \leq n-1 \tag{2.8}$$

Suppose $t = 1$. (2.7) and (2.8) then yield $g_1(y_1) = g_1(-y_1) = -g_1(y_1)$. Since Y is a $\mathbb{Q}$-vector space, $g_1 = f_{n-1} = 0$. Suppose $2 \leq t \leq n-1$. We recall that $/v_1/\ldots/v_t/ = (v_1+\ldots+v_t) + /-v_t/\ldots/-v_1/$. Combining this fact with (2.8) and Proposition 2.2.1, we obtain:

$$2g_t(y_1, \ldots, y_t) = \Sigma_{1\leq i\leq t-1} g_t(/y_i/\ldots/y_1/\#/y_{i+1}/\ldots/y_t/) \tag{2.9}$$

According to the second canonical decomposition, $[m \circ /y_i/\ldots/y_1/]$ is a polynomial in m of degree i without constant term. We can therefore conclude from (2.9) that $g_t(my_1, \ldots, my_t)$ is a polynomial in m of degree at most t without either constant or linear term. As indicated before, this shows that $f_n = 0$. In passing, we note that $f_{n-1} = 0$ in the present case. In any event, we have concluded the first step.

Since f_n is always 0, the general assertion is valid when $n = 2$. We may assume $n > 2$. According to conditions $(\mathfrak{F}_1)$ and $(\mathfrak{F}_2)$ on $\mathfrak{F}(j, \eta)$, we may conclude from Lemma 2.1 that:

$$\mathfrak{a}_{j+1} \cdots \mathfrak{a}_n f_j = 0, \quad 1 \leq j < n$$

Combining this with $f_n = 0$ and $\Sigma_j f_j = 0$, we see that:

$$\mathfrak{a}_3 \cdots \mathfrak{a}_n f_j = 0 \text{ for } 1 \leq j \leq n$$

Let $s \leq n$ be maximal so that:

$$\mathfrak{a}_s \cdots \mathfrak{a}_n f_j = 0 \text{ for } 1 \leq j \leq n \tag{2.10}$$

There are now two cases.

Case 1. $s = n$. Using conditions $(\mathfrak{F}_1)$, $(\mathfrak{F}_2)$ and $(\mathfrak{F}_3)$ on $\mathfrak{F}(j, \eta)$, we can deduce from Proposition 2.2.1 and Lemma 2.2 that:

$$\begin{aligned} f_j(2^s x_1, \ldots, 2^s x_n) &= 2^s f_j(x_1, \ldots, x_j, 2^s x_{j+1}, \ldots, 2^s x_n) \\ &= 2^{s(n-j+1)} f_j(x_1, \ldots, x_n), \quad 1 \leq j \leq n, \ s \in \mathbb{Z} \end{aligned}$$

The assumption $\Sigma_j f_j = 0$ now yields:

$$\Sigma_{1 \leq j \leq n} 2^{s(n-j+1)} f_j = 0, \quad s \in \mathbb{Z} \text{ is arbitrary} \tag{2.11}$$

If we let s range over n distinct integers and use the nonvanishing of the van der Monde determinant of the coefficient matrix in the system of equations obtained from (2.11), then $f_j = 0$ must hold for $1 \leq j \leq n$. We have finished the proof of Lemma 2.5 in this case.

Case 2. $s < n$. We will derive a contradiction to (2.10) by showing:

$$\mathfrak{a}_{s+1} \cdots \mathfrak{a}_n f_j = 0, \quad 1 \leq j \leq n.$$

Since $f_n = 0$ and $\mathfrak{a}_{j+1} \cdots \mathfrak{a}_n f_j = 0$ for $1 \leq j < n$, the desired contradiction is reached if we can show:

$$\mathfrak{a}_{s+1} \cdots \mathfrak{a}_n f_j = 0, \quad 1 \leq j \leq s-1 \tag{2.12}$$

Let $g_j = \mathfrak{a}_{s+1} \cdots \mathfrak{a}_n f_j$, $1 \leq j \leq s-1$. The same two facts plus $\Sigma_j f_j = 0$ yield:

$$\Sigma_{1 \leq j \leq s-1} g_j = 0 \tag{2.13}$$

We will now view $g_j(x_1, \ldots, x_n) = \mathfrak{a}_{s+1} \cdots \mathfrak{a}_n f(x_1, \ldots, x_n)$ as a function of $x_1, \ldots, x_{s-1}$ so that $g_j \in \mathrm{Map}(GL_k(\mathfrak{n}), Y)$ and $(x_1, \ldots, x_{j-1})$ varies over k-bases of an arbitrary but fixed k-subspace $\mathfrak{n}$ complementary to the k-subspace $\mathrm{Lsp}(/x_s/\ldots/x_n/)$ of $\mathfrak{m}$. We assert that $g_j \in \mathfrak{F}(j, -\eta)$. Using the assumption (2.10), we can apply Lemma 2.3 to g_j. Properties $(\mathfrak{F}_1)$, $(\mathfrak{F}_2)$ and $(\mathfrak{F}_3)$ can be verified for $g_1, \ldots, g_{s-1}$ by using the corresponding properties for $f_1, \ldots, f_{s-1}$. We note that these properties have nothing to do with η. Property $(\mathfrak{F}_4)$ is a restriction on $g_1, \ldots, g_{s-2}$. Let $1 \leq j \leq s-2$. (2.10) together with the definition of g_j yields: $\mathfrak{a}_s g_j = 0$, $1 \leq j \leq s-2$. This in turn yields the first equality in the following equations:

$$\begin{aligned} & g_j(x_1, \ldots, x_j, -x_{j+1}, \ldots, -x_{s-1}, x_s, \ldots, x_n) \\ &= g_j(x_1, \ldots, x_j, -x_{j+1}, \ldots, -x_s, x_{s+1}, \ldots, x_n) \\ &= (-1)^{n-s} g_j(x_1, \ldots, x_j, -x_{j+1}, \ldots, -x_s, -x_{s+1}, \ldots, -x_n) \\ &= (-1)^{n-s} \eta (-1)^{n-j-1} g_j(x_1, \ldots, x_n) \\ &= (-\eta)(-1)^{(s-1)-j-1} g_j(x_1, \ldots, x_n) \end{aligned} \tag{2.14}$$

The second equality in (2.14) is a consequence of the skewsymmetry of $g_j = \mathfrak{a}_{s+1} \ldots \mathfrak{a}_n f_j$ in the variables $x_{s+1}, \ldots, x_n$. The third equality in (2.14) follows from $(\mathfrak{F}_4)$ for f_j, $1 \leq j \leq s-2 < n$. The last equality is obvious. Since $s-1 < n$, we conclude from induction that $g_j = 0$, $1 \leq j \leq s-1$. The desired contradiction (2.12) is reached. Q.E.D.

In view of Lemmas 2.4 and 2.5, Theorem 1.1 evidently follows from Lemma 1.2.

3. k-VECTOR SPACE STRUCTURE ON $\mathcal{G}_i(\mathcal{P}^n(G))$.

Let G be any subgroup between $T_k(\mathfrak{m})$ and $A_k(\mathfrak{m})$. Let $1 \leq i \leq n$ and assume that k is i-th root closed (this restriction will be removed later). We will use Theorem 1.1 to extend the $\mathbb{Q}$-vector space structure on $\mathcal{G}_i(\mathcal{P}^n(G))$ to a k-vector space structure. Under the extended definition, scaling by $\lambda \in k^+$ on $\mathfrak{m}$ will correspond to scalar multiplication by λ^i on the k-vector space $\mathcal{G}_i(\mathcal{P}^n(G))$. The assumption on k brings out the key role played by the homotheties $k^+ I_n$; it also brings out the nonlinear nature of scaling on the level of $\mathcal{P}^n(G)$.

Theorem 3.1. Let $\mathfrak{m}$ be a vector space of dimension n over an ordered field k. Assume k is i-th root closed for some i with $1 \leq i \leq n$. For $\lambda \in k^+$ and for any basic i-fold cylinder A in $\mathfrak{m}$, define $\lambda * [A]$ to be $[\lambda^{1/i} \circ A]$. Then:

(a) the $*$ action induces a unique extension of the $\mathbb{Q}$-vector space structure of $\mathcal{G}_i(\mathcal{P}^n(G)) \cong \mathcal{P}^n_i(G)/\mathcal{P}^n_{i+1}(G)$ to that of a k-vector space;

(b) if $\mathfrak{m}$ is the internal direct sum of k-subspaces $\mathfrak{m}_j$ of dimension $d(j) > 0$, $1 \leq j \leq i$ and if G_j is the subgroup of G stablizing V_j and inducing the identity on V_t, $t \neq j$, then the Minkowski sum induces a k-linear map from $\otimes_k \mathcal{G}_1(\mathcal{P}^{d(j)}(G_j))$ to $\mathcal{G}_i(\mathcal{P}^n(G))$.

Proof. Set $0 * [A] = 0$. For $\lambda \in k^-$, define $\lambda * [A]$ to be $-((-\lambda) * [A])$. This extends the $*$ action from k^+ to k. Since scaling is distributive with respect to interior disjoint union, it is straightforward to see that the $*$ action of k is uniquely extendable to all of $\mathcal{P}^n_i(G)$. Except for additivity in λ, the other k-vector space axioms are straightforward. We note that the checking process requires a case analysis of the possibilities that the scalar multipliers may be positive, negative or zero, compare Jessen, Karpf and Thorup [43]. The additivity in λ is generally not valid on the level of $\mathcal{P}_i^n(G)$. However, it becomes valid when we pass to $\mathcal{G}_i(\mathcal{P}^n(G))$ via congruence mod $\mathcal{P}_{i+1}^n(G)$. We first show:

$$[(\lambda+\mu)^{1/i} \circ A] \equiv [\lambda^{1/i} \circ A] + [\mu^{1/i} \circ A] \bmod \mathcal{P}_{i+1}^n(G) \tag{3.2}$$

where $\lambda, \mu \in k^+$ and A is a basic i-fold cylinder.

When i = 1, (3.2) follows from the first canonical decomposition. When i > 1, Theorem 1.1 and induction together imply:

$$\begin{aligned}[\lambda^{1/i} \circ A] &= [(\lambda^{1/i} \circ A_1) \# \ldots \# (\lambda^{1/i} \circ A_i)] \\ &\equiv [A_1 \# \ldots \# (\lambda \circ A_j) \# \ldots \# A_i] \bmod \mathcal{P}_{i+1}^n(G), \; 1 \le j \le i\end{aligned} \tag{3.3}$$

(3.2) follows from (3.3) together with first canonical decomposition. The general distributive law follows from (3.2) after a case analysis. We leave the details to the patient reader. Assertion (a) now holds. Assertion (b) follows from (a) and Theorem 1.1. Q.E.D.

Theorem 3.4. Let $\mathfrak{M}$ be a vector space of dimension n over an ordered field k. Assume k is n!-th root closed and assume G is a subgroup between $T_k(\mathfrak{M})$ and $A_k(\mathfrak{M})$ so that G fixes a positive definite inner product $\langle\, , \rangle$ on $\mathfrak{M}$. $\mathcal{P}^n(G)$ then has a unique k-vector space structure with the following properties:

(a) $\mathcal{P}^n(G)$ is a k-vector space direct sum of the k-subspaces $\mathcal{G}_i^n(G)$;

(b) scaling by $\lambda \in k^+$ on $\mathfrak{M}$ induces scalar multiplication by λ^i on $\mathcal{G}_i^n(G)$;

(c) the Hadwiger invariant $\Omega^n(r_1, \ldots, r_j)$ is an element of k^+-weight n-j in $\mathrm{Hom}_k(\mathcal{P}^n(T_k(\mathfrak{M})), k)$ when $G = T_k(\mathfrak{M})$.

The proof is straightforward and will be omitted.

4 Completeness of Hadwiger invariants

Let K be any ordered, square root closed field. It serves the purpose of a universal domain. Let $\mathfrak{M}$ be a vector space of dimension n over a subfield k of K. Let < , > be a k-valued positive definite inner product on $\mathfrak{M}$. After ground field extension to K, any other choice becomes equivalent to < , >. The main goal of the present chapter is to prove the following assertion:

> Two polyhedra P and Q in $\mathfrak{M}$ are stably scissors congruent with respect to the group $T_k(\mathfrak{M})$ of all translations if and only if they have the same set of Hadwiger invariants, i.e., if and only if:
>
> $\Omega^n(r_1, \ldots, r_j)(P) = \Omega^n(r_1, \ldots, r_j)(Q)$ holds for all j-frames $(r_1, \ldots, r_j)$ of $\mathfrak{M}$. $0 \leq j \leq n-1$. (Recall that j = 0 corresponds to volume.)

If k is Archimedean, then the word "stably" may be removed by using the theorem of Zylev (which, in the present case, was proved earlier by H. Hadwiger). For n = 1, the assertion is trivial. For n = 2, it is due to Hadwiger-Glur [34]. For n = 3, it is due to Hadwiger [33]. For general n, it is first proved by Jessen-Thorup in 1972-73, see [44]. Without being aware of this fact, we gave another proof in 1977 [60] under the easily removed assumption that k be Archimedean. The present proof follows our 1977 account because it eases our way into the following chapter when we study the syzygy problem. Unless stated explicitly, the group G of motions of $\mathfrak{M}$ is taken to be the group $T_k(\mathfrak{M})$ of all translations in the present chapter. For this reason, it will usually be omitted. In general, $\mathfrak{P}^n$ and $\mathfrak{G}_i^n$ are $\mathbb{Q}$-vector spaces with scaling action by k^+. Except for $\mathfrak{G}_1^n$, they are not yet k-vector spaces. In fact, one of the corollaries of the main result will be the validity of Theorem 3.3.4 without the assumption of n!-th root closure of k. Ultimately, $\mathfrak{P}^n$ and $\mathfrak{G}_i^n$ will be k-vector spaces.

1. PRELIMINARIES ON THE MAIN THEOREM.

Theorem 1.1. Let K be an ordered, square root closed field. Let $\mathfrak{M}$ be a vector space of dimension n over the subfield k of K. Let < , > be a positive definite k-valued inner product on $\mathfrak{M}$. Let $1 \leq i \leq n$. Then the set of all Hadwiger invariants $\Omega^n(r_1, \ldots, r_{n-i})$ with $r_1, \ldots, r_{n-i}$ ranging over all (n-i)-frames in $\mathfrak{M}$ separates the points of $\mathcal{G}_i^n = \mathcal{G}_i(\mathcal{P}^n(T_k(\mathfrak{M})))$ (we will identify $\mathcal{G}_i^n$ with $\mathcal{P}_i^n / \mathcal{P}_{i+1}^n$ as in Proposition 2.5.1).

Suppose i = n. Theorem 1.1 follows from Theorem 2.5.1. We will assume $1 \leq i \leq n-1$ so that n > 1. The general proof will be carried out by induction. We first consider the case of $\mathcal{G}_1^n$. Here we take advantage of the fact that $\mathcal{G}_1^n$ is a k-vector space. The concluding step will then be carried out under the added assumptions that 1 < i < n and that Theorem 1.1 has been verified for all $\mathcal{G}_j^m$ if either m < n or j < i. For this second step, we do not assume that $\mathcal{G}_i^n$ is a k-vector space when 1 < i < n. However, we do utilize the scaling action by k^+. Before going on with the proofs, we introduce some terminologies and state a preliminary result.

Let V be an arbitrary k-vector space. An indexed set of rays $\{r_j | j \in J\}$ in V will be called k-independent when the following condition holds:

> If $y_j \in \pm r_j \cup \{0\}$ with $y_j = 0$ for almost all y_j, then $\Sigma_j y_j = 0$
> if and only if $y_j = 0$ for every j in J.

If we select v_j on r_j for each j in J, then $\{r_j | j \in J\}$ is k-independent if and only if $\{v_j | j \in J\}$ is k-independent as an indexed set of vectors in V.

Fix a ray r_0 in $\mathfrak{M}$. An (s-1)-frame $r_1, \ldots, r_{s-1}$ will be called $\underline{r_0}$-positive when the following conditions hold:

(a) $r_0, r_1, \ldots, r_{s-1}$ are k-independent; and
(b) $< r_0, r_i > = k^+$ for $1 \leq i \leq s-1$.

An (s-1)-frame $r_1, \ldots, r_{s-1}$ will be called $\underline{r_0}$-negative if $-r_1, \ldots, -r_s$ is r_0-positive. An orthogonal n-simplex $z + /y_1/\ldots/y_n/$ is called $\underline{r_0}$-positive (respectively, $\underline{r_0}$-negative) if $r_0 = k^+ y_1$ (respectively, $r_0 = -k^+ y_1 = k^- y_1$). In general, the orthogonal n-simplex $z+/y_1/\ldots/y_n/$ has exactly one other representation exhibiting the orthogonality: $(z+y_1+\ldots+y_n)+/-y_n/\ldots/-y_1/$. It follows that r_0-positivity and r_0-negativity are exclusive concepts for a geometric n-simplex when $n \geq 2$. When the property holds, the nonhomogeneous representation exhibiting the property is unique.

More generally, an orthogonal basic i-fold cylinder $A = A_1\#\ldots\#A_i$ will be called r_0-positive (respectively r_0-negative) if some A_j (j is then unique) is r_0-positive (respectively r_0-negative) in $\mathrm{Lsp}(A_j)$ after a translation. Permuting $A_1, \ldots, A_i$ when necessary, we agree to take $j = 1$. In all cases, we extend the definition to the class of A in $\mathcal{P}^n(T_k(\mathfrak{m}))$.

Lemma 1.2. Let r_0 be a fixed ray in $\mathfrak{m}$. Let $1 \leq i \leq n-1$. Then each element of $\mathcal{C}_i^n$ is represented by a finite sum (no subtraction needed) of the classes of r_0-negative orthogonal basic i-fold cylinders.

The proof follows from Theorem 2.2.4 and Proposition 2.5.4 via an easy adaptation of their proofs. We leave the details to the patient reader. It should be noted that r_0-positive orthogonal basic i-fold cylinders are not needed because of the following identity:

> $\chi(\sigma A) = (-1)^{n-i}\chi(A)$, $\chi \in \mathrm{Hom}_{\mathbb{Z}}(\mathcal{C}_i^n, \mathbb{Q})$, $\sigma = -I_n$ and A is any orthogonal basic i-fold cylinder.

2. BEGINNING OF THE PROOF.

Let B be a subset of the k-vector space V. Let B^* be a subset of the K-dual $V^* = \mathrm{Hom}_k(V, K)$. B and B^* are said to be bases in ray duality if the following conditions hold:

(a) B is a k-basis for V; and

(b) there is a bijection $\nu: B \longrightarrow B^*$ such that $(\nu(b))(b) > 0$ for b in B and $(\nu(a))(b) = 0$ for $a \neq b$ in B.

The uniqueness of ν is an immediate consequence of (b). B^* is not uniquely determined by B, however, the set of rays spanned by elements of B^* is uniquely determined by B. The same holds with B and B^* interchanged. In general B^* is not a K-basis for V^* (it is so if and only if $\dim_k V$ is finite). It is clear that any subset between B^* and V^* will separate the points of V.

Let r_0 be a fixed ray in $\mathfrak{M}$. Let J be the set of all r_0-positive (n-1)-frames in $\mathfrak{M}$. Let B* be the set of all Hadwiger invariants $\Omega^n(r_1, \ldots, r_{n-1})$ with $(r_1, \ldots, r_{n-1})$ in J. The set of r_0-negative orthogonal n-simplices of the form $/y_1/\ldots/y_n/$ is clearly stable under the scaling action by k^+. Let B be the set of all elements of $\mathfrak{G}^n_1$ represented by [A] where A ranges over the distinct k^+-orbits of r_0-negative orthogonal n-simplices of the form $/y_1/\ldots/y_n/$. The first step of the proof is then a consequence of the following stronger result:

Lemma 2.1. B and B* are bases in ray duality for the k-vector space $\mathfrak{G}^n_1$ and the K-dual $\mathrm{Hom}_k(\mathfrak{G}^n_1, K)$.

Proof. It is clear that B* is indexed by J. The duality map ν is found by the description of a map from B to J.

Let $A = /y_1/\ldots/y_n/$ be a r_0-negative orthogonal n-simplex in $\mathfrak{M}$. We form the following codimensional 1 face sequence:

$$/y_1/\ldots/y_n/,\ /y_1+y_2/y_3/\ldots/y_n/,\ \ldots\ ,\ /y_1+\ldots+y_n/ \tag{2.2}$$

Let $r_1, \ldots, r_{n-1}$ be the associated sequence of exterior normal rays. It is clearly an (n-1)-frame depending only on the k^+-orbit of A. We will have a candidate for the duality map ν as soon as we show that $(r_1, \ldots, r_{n-1})$ is r_0-positive. It is immediate that:

$$\mathrm{Lsp}(r_0, r_1, \ldots, r_{s-1}) = \mathrm{Lsp}/y_1/\ldots/y_s/,\ 1 \leq s \leq n \tag{2.3}$$

According to definition, r_0-positivity of $(r_1, \ldots, r_{n-1})$ is equivalent with (2.3) and the following:

$$< y_1, r_i > = -< r_0, r_i > = -k^+ = k^-,\ 1 \leq i \leq n-1 \tag{2.4}$$

A vector x_i in $\mathrm{Lsp}(y_1+\ldots+y_i, y_{i+1})$ exterior normal to the face $/(y_1+\ldots+y_i)+y_{i+1}/$ of $/y_1+\ldots+y_i/y_{i+1}/$ must have the form:

$$-\alpha_i(y_1+\ldots+y_i) + \beta_i y_{i+1},\ \alpha_i,\ \beta_i \in k^+ \tag{2.5}$$

Since $/y_1/\ldots/y_n/$ is orthogonal, $< x_i, y_1 > = -\alpha_i < y_1, y_1 > \in k^-$. This proves (2.4) so that $(r_1, \ldots, r_{n-1}) \in J$ as desired. (see diagram (*).)

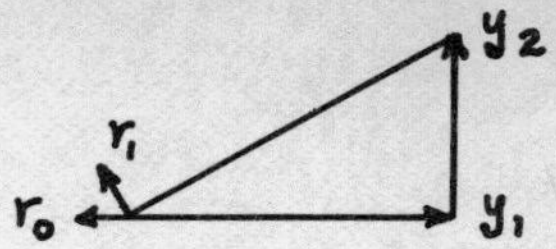

(*)

We now have a map ν: B ⟶ B*. It is clear that $\Omega^n(r_1, \ldots, r_{n-1})$ takes the value $\|y_1 + \ldots + y_n\|^2 \in K^+$ on $/y_1/\ldots/y_n/$ where $\|v\|^2 = \langle v, v \rangle$. Thus, $(\nu(b))(b) \in K^+$ for every b in B. From Lemma 2.1 and the structure of $\mathcal{G}_1^n$ as a k-vector space, B generates $\mathcal{G}_1^n$ as a k-vector space. The map ν is then a ray duality as asserted as soon as we show:

> For any $(r_1, \ldots, r_{n-1}) \in J$, there is a unique r_0-negative orthogonal n-simplex $/y_1/\ldots/y_n/$ (unique up to k^+-scaling) with the property: $\Omega^n(r_1, \ldots, r_n)[/y_1/\ldots/y_n/] \neq 0$. When this is the case, $(r_1, \ldots, r_{n-1})$ is the sequence of exterior normal rays associated to the codimensional 1 face sequence described by (2.2). (2.6)

Let $x_i \in r_i$, $1 \leq i \leq n-1$. An examination of the argument so far shows that (2.6) amounts to finding the orthogonal k-basis $y_1, \ldots, y_n$ with y_1 in $-r_0$ in terms of $x_1, \ldots, x_{n-1}$ so that the answer is unique up to scaling action of k^+. We will now verify (2.6) by induction on n.

We begin with an arbitrary r_0-negative orthogonal n-simplex. The codimensional 1 faces are:

$$y_1+/y_2/\ldots/y_n/, \ /y_1+y_2/\ldots/y_n/, \ldots$$
$$/y_1/\ldots/y_i+y_{i+1}/\ldots/y_n/, \ldots, /y_1/\ldots/y_{n-1}/$$

With the exception of $/y_1+y_2/y_3/\ldots/y_n/$, the exterior normal rays to the remaining faces must be $\pm r_0$ or be orthogonal to r_0. In order for r_1 to appear as a normal ray to one of these codimensional 1 faces and still satisfy the condition $(r_1, \ldots, r_{n-1}) \in J$, we must select $/y_1+y_2/y_3/\ldots/y_n/$. It then follows that r_1 is the exterior normal (see diagram (*) above). The geometric picture in diagram (*) easily translates to a problem in inner product spaces over an ordered field. If $y_1 \in -r_0$ is selected at random (this accounts for uniqueness up to k^+-scaling action), then y_2 is uniquely determined in $Lsp(r_0, r_1) = ky_1 + kx_1 = ky_1 + ky_2$. According to definition:

$$\Omega^n(r_1, \ldots, r_{n-1})[/y_1/\ldots/y_n/] = \Omega^{n-1}(r_2, \ldots, r_{n-1})[/y_1+y_2/\ldots/y_n/]$$

If we let $\mathfrak{h} = Lsp/y_1+y_2/y_3/\ldots/y_n/$ be the orthogonal complement of r_1 and let $s_0 = -k^+(y_1+y_2)$, then it is easy to check that $(r_2, \ldots, r_{n-1})$ is an s_0-positive (n-2)-frame in $\mathfrak{h}$. Since $y_1+y_2 \in -s_0$ is specified, we may deduce from induction the existence and uniqueness of an orthogonal (n-1)-simplex $/y_1+y_2/y_3/\ldots/y_n/$ in $\mathfrak{h}$ so that:

> $\Omega^{n-1}(r_2, \ldots, r_{n-1})[/y_1+y_2/y_3/\ldots/y_n/] \neq 0$ and with $(r_2, \ldots, r_{n-1})$ as the sequence of exterior normal rays described in (2.2).

It is now evident that (2.6) holds. Q.E.D.

3. CONCLUSION OF THE PROOF.

Let $A = A_1 \# \ldots \# A_i$ be any orthogonal basic i-fold cylinder, $1 < i < n$. We want to evaluate a Hadwiger invariant $\Omega^n(r_1, \ldots, r_{n-j})$ on A. In order to obtain a nonzero value, at least one codimensional 1 face of A must have r_1 as a normal ray. A typical codimensional 1 face of A has the form:

$$F(A) = A_1 \# \ldots \# A_{s-1} \# F(A_s) \# A_{s+1} \# \ldots \# A_i$$

where $F(A_s)$ is a suitable codimensional 1 face of A_s, $1 \leq s \leq i$. Unless $\dim_k Lsp(A_s) = d(s)$ is 1, F(A) is an orthogonal basic i-fold cylinder of dimension n-1. When $d(s) = 1$, F(A) is an orthogonal basic (i-1)-fold cylinder of dimension n-1. If $d(s) > 1$, F(A) is uniquely determined by its exterior normal ray. This normal is just the exterior normal ray to the face $F(A_s)$ with respect to A_s in $Lsp(A_s)$. In case $d(s) = 1$, $A_s = /x_s/$ up to translation and we have a pair of faces which are congruent under a translation by x_s. The interior normal ray of one is the exterior normal ray of the other. As a result, we have cancellation of the values of the Hadwiger invariant $\Omega^n(r_1, \ldots, r_{n-j})$. This analysis is clearly inductive. If we also recall the weight space decomposition, then the analysis can be summarized in the following assertion:

Lemma 3.1. Let $r_1, \ldots, r_{n-j}$ be an (n-j)-frame in $\mathfrak{M}$, $1 \leq j \leq n-1$. Let $A = A_1\#\ldots\#A_i$ be an orthogonal basic i-fold cylinder in $\mathfrak{M}$. $\Omega^n(r_1, \ldots, r_{n-j})$ is nonzero on A if and only if: $i \leq j$ and there is a unique index s with the following properties:

(a) $\dim_k \operatorname{Lsp}(A_s) > 1$;

(b) $\epsilon_1 r_1$ is the exterior normal ray to a uniquely determined codimensional 1 face $F(A_s)$ of A_s with respect to A_s, $\epsilon_1 = \pm 1$;

(c) $\Omega^{n-1}(r_2, \ldots, r_{n-j})$ is nonzero on $A_1\#\ldots\#F(A_s)\#\ldots\#A_i$.

When all these happen, we have:

(d) $\Omega^n(r_1, \ldots, r_{n-j})[A]$

$$= \epsilon_1 \Omega^{n-1}(r_2, \ldots, r_{n-j})[A_1\#\ldots\#F(A_s)\#\ldots\#A_i].$$

Lemma 3.2. Let $1 < i < n$. Assume Theorem 1.1 has been verified for $\mathcal{G}^m_j$ when either $m < n$ or $j < i$. Then Theorem 1.1 holds for $\mathcal{G}^n_i$.

Proof. Let $A = A_1\#\ldots\#A_i$ be an r_0-negative orthogonal basic i-fold cylinder in $\mathfrak{M}$. By agreement, we assume A_1 is an r_0-negative orthogonal d(1)-simples in $\operatorname{Lsp}(A_1)$ with $d(j) = \dim_k \operatorname{Lsp}(A_j) > 0$. In all cases (allowing i to be between 1 and n), the ray spanned by A in defined to be the set:

$$\{[(\lambda \circ A_1)\#A_2\#\ldots\#A_i \bmod \mathcal{P}^n_{i+1} \mid \lambda \in k^+\} \tag{3.3}$$

According to Theorem 3.1.1, the set defined by (3.3) is identical with:

$$\{[A_1\#\ldots\#(\lambda \circ A_j)\#\ldots\#A_i \bmod \mathcal{P}^n_{i+1} \mid \lambda \in k^+\}, \; 1 \leq j \leq i \tag{3.4}$$

These rays are viewed as subsets of $\mathcal{G}^n_i$ through the canonical isomorphism $\mathcal{G}^n_i \cong \mathcal{P}^n_i/\mathcal{P}^n_{i+1}$. According to Lemma 1.2, each element $[B]$ of $\mathcal{G}^n_i$ is a finite sum of elements belonging to these rays. Our assertion is equivalent with the following statement:

> If $[B]$ is a finite sum of elements in rays described by (3.4), then $[B] \equiv 0 \bmod \mathcal{P}^n_{i+1}$ holds if and only if $\Omega^n(r_1, \ldots, r_{n-i})[B] = 0$ holds for all (n-i)-frames $r_1, \ldots, r_{n-i}$ in $\mathfrak{M}$.

The necessity follows from weight consideration. We therefore assume that $\Omega^n(r_1, \ldots, r_{n-i})[B] = 0$ for all (n-i)-frames $r_1, \ldots, r_{n-i}$ in $\mathfrak{m}$.

Let $\mathfrak{n}$ be a k-subspace of $\mathfrak{m}$ with the following properties:

$$r_0 \subset \mathfrak{n} \text{ and } \dim_k \mathfrak{n} \leq n-i+1 \text{ (note: } 2 \leq n-i+1 \leq n-1) \tag{3.5}$$

If $A = A_1 \# \ldots \# A_i$ is any r_0-negative, orthogonal, basic i-fold cylinder in $\mathfrak{m}$ and $r_0 \subset \mathrm{Lsp}(A_1)$, then $\mathrm{Lsp}(A_1)$ satisfies (3.5). It is clear that our $[B]$ can be written as a finite sum:

$[B] = \Sigma[B(U)]$, U ranging over a finite set satisfying (3.5)

and,

$[B(U)]$ is a finite sum of $[A]$ where $A = A_1 \# \ldots \# A_i$ is a r_0-negative, orthogonal, basic i-fold cylinder in $\mathfrak{m}$ with $\mathrm{Lsp}(A_1) = U$, U satisfies (3.5). (3.6)

The desired assertion clearly follows if we show:

Under the assumption on $[B]$, $[B(U)] \equiv 0 \bmod \mathcal{P}^n_{i+1}$ holds for each U satisfying (3.5) (3.7)

In view of (3.3) and (3.4), each $[B(U)]$ is a finite sum:

$[B(U)] = \Sigma_j [B_j \# C_j]$, where B_j ranges over distinct rays of r_0-negative orthogonal $\dim_k U$-simplices in U and C_j is an interior disjoint union of orthogonal, basic (i-1)-fold cylinders in the orthogonal complement U' of U. (3.8)

By way of contradiction, assume (3.7) does not hold. We can therefore find U of maximal dimension $s \leq n-i+1$ so that (3.7) does not hold. Consider (n-i)-rays $r_1, \ldots, r_{n-i}$ with the following properties:

(a) $r_0, r_1, \ldots, r_{s-1}$ are k-independent;
(b) $< r_j, r_0 > = k^+$ for $1 \leq j \leq s-1$; and (3.9)
(c) $< r_j, r_0 > = 0$ for $s \leq j \leq n-i$.

We now evaluate $\Omega^n(r_1, \ldots, r_{n-i})$ on $[B(W)]$, W satisfying (3.5) as before. If $\dim_k(W) < s$, then Lemma 4.1 shows that we must get 0 on each of the terms making up $[B(W)]$ in (3.6). If $\dim_k W = s$, and $W \neq U$, then Lemma 2.1 and Lemma 3.1 together imply that we get 0. In view of the assumption on $[B(U)]$, we automatically get 0 when $\dim_k W > s$. In view of the assumption on $[B]$, we must also get 0 on $[B(U)]$ as long as $(r_1, \ldots, r_{n-i})$ satisfies (3.9). The ray duality in Lemma 2.1, the inductive evaluation process in Lemma 3.1, together with the sum representation in (3.8) for $[B(U)]$ yield:

$$\Omega^{n-s}(r_s, \ldots, r_{n-i})[C_j] = 0 \text{ for each } j, \text{ where } r_s, \ldots, r_{n-i} \text{ is any } \{(n-s)-(i-1)\}\text{-frame in the orthogonal complement } U' = \mathrm{Lsp}(C_j) \text{ of } U. \qquad (3.10)$$

Since $[C_j] \in \mathcal{P}^{n-s}_{i-1}(T_k(U'))$ and $n-s < n$, (3.10) together with the inductive hypotheses in our assertion imply:

$$[C_j] \equiv 0 \bmod \mathcal{P}^{n-s}_i \text{ holds for each } j \qquad (3.11)$$

The desired contradiction now follows from (3.8) and (3.11). Q.E.D.

It is now evident that Theorem 1.1 follows from Lemmas 2.1 and 3.2.

4. k-VECTOR SPACE STRUCTURE ON $\mathcal{P}^n(G)$.

Proposition 4.1. Let $k \subset k'$ be subfields of K. Select a basis and identify n-dimensional vector spaces $\mathfrak{m}$ and $\mathfrak{m}'$ over k and k' with column vectors of length n. The natural inclusion map from $\mathfrak{m}$ to $\mathfrak{m}'$ then induces an injection of $\mathcal{P}^n(\mathfrak{m})$ into $\mathcal{P}^n(\mathfrak{m}')$.

Proof. Since all inclusion maps are compatible, we may take $k' = K$ and identify $\mathfrak{m}$ and $\mathfrak{m}'$ with k^n and K^n respectively. The positive definite inner product $<\ ,\ >$ can be taken to be the usual one. As a result, any j-frame in k^n is automatically a j-frame in K^n. The injectivity assertion is equivalent with the following:

$$\mathrm{im}\ \mathcal{P}^n_i(T_k(k^n)) \cap \mathcal{P}^n_{i+1}(T_K(K^n)) \subset \mathrm{im}\ \mathcal{P}^n_{i+1}(T_k(k^n)),\quad 1 \leq i \leq n.$$

In view of Theorem 2.2.5, the assertion is clear when $i = n$. For $i < n$, the assertion follows from a combination of Propositions 2.2.6, 2.5.3 and Theorem 1.1. Q.E.D.

We are now ready to remove the n!-th root closure assumption placed on k in Theorem 3.3.4. For $0 \leq i \leq n-1$, let $St_i^n(k)$ denote the Stiefel variety of the set of all i-frames in $\mathfrak{M}$ (when k is itself square root closed, we use the unique unit vector on each ray in place of the ray itself). By agreement, $St_0^n(k)$ is understood to be a point. $Map(St_i^n(k), K)$ is then a K-vector space through the pointwise action of K on the values of each map. Through the process of evaluation of the Hadwiger invariants, each element of $\mathcal{P}^n(T_k(\mathfrak{M}))$ defines an element of $\coprod_i Map(St_i^n(k), K)$. This map from $\mathcal{P}^n(T_k(\mathfrak{M}))$ to the K-vector space $\coprod_i Map(St_i^n(k), K)$ will be called the <u>Hadwiger map</u> and will be denoted by h. The direct sum decomposition displayed allows us to write $h = \Sigma_i h_i$, $0 \leq i \leq n-1$. If P is any polyhedron in $\mathfrak{M}$, then:

$$h_i[P](r_1, \ldots, r_i) = \Omega^n(r_1, \ldots, r_i)[P]$$

<u>Theorem 4.2.</u> With the preceding notations, the Hadwiger map h is an injection. The image of h is a k-vector subspace of the space of all finitely supported maps on the set $\cup_i St_i^n(k)$ (disjoint union) with values in K. When k is n!-th root closed, h is a k-vector space injection with respect to the k-vector space structure on $\mathcal{P}^n(T_k(\mathfrak{M}))$ described in Theorem 3.3.4.

<u>Proof</u>. The map h is clearly additive. The injectivity of h is just a restatement of Theorem 1.1. In general, a geometric n-simplex has only a finite number of distinct codimensional 1 face sequences of a prescribed length i. It is therefore clear that the image of h lies in the space of all finitely supported maps on $\cup_i St_i^n(k)$. In order to show that im h is a k-subspace, it is enough to show that a set of generators of im h (known to be an additive subgroup) is closed with respect to multiplication by elements of k^+. For this, we may restrict ourselves to each h_{n-i}. By using weight space decomposition and the definition of h_{n-i}, im h_{n-i} is generated by the set of all $h_{n-i}[A]$ with A ranging over orthogonal basic i-fold cylinders. Direct calculation by means of Lemma 3.1 shows that:

$$h_{n-i}(A_1 \# \ldots \# (\lambda \circ A_j) \# \ldots \# A_i) = \lambda h_{n-i}(A_1 \# \ldots \# A_i)$$

$A_1 \# \ldots \# A_i$ any orthogonal basic i-fold cylinder in $\mathfrak{M}$.

Q.E.D.

From now on, $\mathcal{P}^n(T_k(\mathfrak{m}))$ will be a k-vector space via the map h.

<u>Theorem 4.3.</u> Let G be any subgroup between $T_k(\mathfrak{m})$ and the group $E_k(\mathfrak{m}) = E_k(\mathfrak{m}, < , >)$ of all rigid motions of $\mathfrak{m}$. Then:

(a) the Hadwiger map h is G-equivariant;

(b) $\mathcal{P}^n(G)$ has a natural k-vector space structure as a quotient of $\mathcal{P}^n(T_k(\mathfrak{m}))$; indeed, $\mathcal{P}^n(G) \cong H_0(G/T_k(\mathfrak{m}), \mathcal{P}^n(T_k(\mathfrak{m})))$, where the latter denotes the 0-th Eilenberg-MacLane homology group of $G/T_k(\mathfrak{m})$ in the left $G/T_k(\mathfrak{m})$-module $\mathcal{P}^n(T_k(\mathfrak{m}))$ with the action defined through the action of G on geometric n-simplices in $\mathfrak{m}$;

(c) $\dim_k \mathfrak{G}_n(\mathcal{P}^n(G)) = 1$;

(d) if $1 \leq i < n$, then $\dim_k \mathfrak{G}_i(T_k(\mathfrak{m}))$ equals to the cardinality of k;

(e) if $-I_n \in G$ and $n-i \equiv 1 \bmod 2$, then $\mathfrak{G}_i(\mathcal{P}^n(G)) = 0$.

<u>Proof.</u> (a) follows from the definition and the following equation:

$$\Omega^n(\sigma r_1, \ldots, \sigma r_i)[\sigma P] = \Omega^n(r_1, \ldots, r_i)[P], \ \sigma \in E_k(\mathfrak{m})$$

(b) follows from (a) and the wellknown identification of the 0-th Eilenberg-MacLane homology groups. (c) is merely Theorem 2.2.5. (e) follows from Proposition 2.5.5. The upper bound for the dimension in (d) follows from the fact that k is infinite and the number of geometric n-simplices in $\mathfrak{m}$ is equal to the cardinality of k. When $i = 1 < n$, the lower bound follows from Lemma 2.1. When $1 < i < n$, let C be an orthogonal parallelopiped in the orthogonal complement of an $(n-i+1)$-dimensional subspace $\mathfrak{h}$. The discussion preceding Lemma 3.1 easily shows that the Minkowski sum with C induces an injection from $\mathfrak{G}_1^{n-i+1}$ to $\mathfrak{G}_1^n$. We get the lower bound from the case just considered. Q.E.D.

5 Syzygies

In the present chapter, we investigate the image of the Hadwiger map h in Theorem 4.4.2. By duality, this is equivalent with the study of "relations" among the Hadwiger invariants. To simplify the investigation, we consider the "stable" problem. Stability means that we will work over an ordered, square root closed field k. The affine n-space will be taken to be the space k^n of all colume vectors of length n over k. The positive definite inner product $<\ ,\ >$ will be taken to be the standard one. As in the preceding chapter, we will use $\mathcal{P}^n$ and $\mathcal{G}_i^n$ in place of $\mathcal{P}^n(T_k(k^n))$ and $\mathcal{G}_i(\mathcal{P}^n(T_k(k^n)))$. The Hadwiger invariants will be based on i-frames of unit vectors rather than rays. $St_i^n(k)$ will denote the Stiefel variety of all i-frames in k^n. The square root closure assumption on k shows that SO(n;k) is transitive on $St_i^n(k)$ for $0 \leq i < n$. $St_0^n(k)$ will be understood to be a point. The investigation of the relations is carried out in the setting of a syzygy sequence--an idea that goes back to Hilbert. The analysis of special cases of the syzygy sequences together with results described in later chapters lead to the appearance of some low dimensional Eilenberg-MacLane cohomology groups of SO(n;k) with coefficients in suitable exterior powers. This seems to be somewhat similar (though more elementary) to the appearance of such homology (with trivial coefficient) in the work of Cheeger-Simons [19]. We conclude this chapter with some open problems.

1. BASIC RELATIONS.

Let $(u_1, \ldots, u_i) \in St_i^n(k)$ and let $f(u_1, \ldots, u_i) \in k$. For any geometric n-simplex A in k^n, $\Omega^n(u_1, \ldots, u_i)(A)$ is nonzero for all but a finite number of points $(u_1, \ldots, u_i)$ of $St_i^n(k)$, $0 \leq i < n$. As a consequence, the expression:

$$\Sigma\, f(u_1, \ldots, u_i)\Omega^n(u_1, \ldots, u_i)$$

with the summation extending over all $(u_1, \ldots, u_i) \in St_i^n(k)$, actually defines an element of $\mathrm{Hom}_k(\mathcal{G}^n_{n-i}, k)$. This is just the space of all functionals of k^+-weight n-i on $\mathcal{P}^n$, we will abbreviate it to $J^n_{n-i}(k)$ or J^n_{n-i} (the letter J is in honor of Jessen, see Hadwiger [31; p. 53]).

Proposition 1.1. Let k be an ordered, square root closed field. For each $(u_1, \ldots, u_{n-i}) \in \mathrm{St}^n_{n-i}(k)$, $1 \leq i \leq n$, we have:

(R_0) $\quad \Omega^n(\eta_1 u_1, \ldots, \eta_{n-i} u_{n-i}) = \eta_1 \cdot \ldots \cdot \eta_{n-i} \Omega^n(u_1, \ldots, u_{n-i})$ in $J^n_i(k)$, where $\eta_s = \pm 1$.

Proof. (R_0) follows directly from the definition of Hadwiger invariants.

Proposition 1.2. Let k be an ordered, square root closed field. Assume $n > 1$ and let $1 \leq i \leq n-1$. Let $E(n, i+1)$ denote the set of all $(u_1, \ldots, u_{n-i-1}; w)$ with $(u_1, \ldots, u_{n-i-1}) \in \mathrm{St}^n_{n-i-1}(k)$ and with w ranging over the orthogonal complement W of $\mathrm{Lsp}(u_1, \ldots, u_{n-i-1})$ in k^n. For each $(u_1, \ldots, u_{n-i-1}; w)$ in $E(n, i+1)$, we have:

(R_1) $\quad \Sigma_u \langle w, u \rangle \Omega^n(u_1, \ldots, u_{n-i-1}, u) = 0$ in $J^n_i(k)$, where the summation extends over all unit vectors u in W.

Proof. Let $U = \mathrm{Lsp}(u_1, \ldots, u_{n-i-1})$. Let A be any geometric n-simplex in k^n. $\Omega^n(u_1, \ldots, u_{n-i-1}, u)(A)$ is automatically zero unless A has at least one (hence exactly one) (i-1)-dimensional face F that can be reached by a codimensional 1 face sequence with associated exterior unit normals $\eta_s u_s$. In particular, $\mathrm{Lsp}(F) = W$. Combining this observation with the definition of Hadwiger invariants allow us to assume $n-i-1 = 0$. We can also assume $w \neq 0$. By Lemma 4.1.2, we only have to show that the left hand side of (R_1) vanishes on all orthogonal (n-1)-fold cylinders of the form:

$$/w/x_1/ \# /x_2/ \# \ldots \# /x_{n-1}/$$

In effect, we are reduced to the case $n = 2$. If we let $v = -\alpha w + \beta x_1$ be the exterior unit normal to the face $/w+x_1/$ of $/w/x_1/$ in $\mathrm{Lsp}(w, x_1)$, $\alpha, \beta \in k^+$, then the vanishing assertion reduces to the easily checked equation:

$$2 \|x_1\| + 2 \langle w, v \rangle \|w+x_1\| = 0.$$

Q.E.D.

Proposition 1.3. Let k be an ordered, square root closed field. Assume $n > 2$. Let $1 \leq i \leq n-1$ and $1 \leq j \leq n-i-1$. Let $F(n,i,j)$ denote the set of all $(u;W;e(W))$ with $u = (u_1, \ldots, u_{j-1}, u_{j+2}, \ldots, u_{n-i}) \in St^n_{n-i-2}(k)$, W a 2-plane in k^n orthogonal to $Lsp(u)$, and $e(W) = x \wedge y \in \Lambda^2_k(W)$ with $(x,y) \in St^2_2(W)$. For each $(u;W;e(W))$ in $F(n,i,j)$, we have:

(R_2) $\quad \Sigma_{u \wedge v = e(W)} \Omega^n(u_1, \ldots, u_{j-1}, u, v, u_{j+2}, \ldots, u_{n-i}) = 0$ in $J^n_i(k)$ where the summation extends over all 2 frames (u,v) in W with orientation specified by $e(W)$ or by (x,y).

Proof. As in the proof of Proposition 1.2, we may assume $j = 1$. Moreover, it is enough to show that the left hand side of (R_2) vanishes on every orthogonal n-simplex A of the form $/z_1/z_2/\ldots/z_n/$ with $W = Lsp(z_1, z_2)$. There are exactly 2 codimensional 1 faces A_1 and A_2 of A containing the codimensional 2 face $F = /z_3/\ldots/z_n/$ of A. The associated exterior unit normals then lead to 2 frames (x_s, y_s) in W with:

$$x_1 \wedge y_1 = -x_2 \wedge y_2 = \epsilon(x \wedge y) = \epsilon e(W), \quad \epsilon = \pm 1$$

The evaluation of the left hand side of (R_2) on A yields:

$$\Omega^{n-2}(u_3, \ldots, u_{n-i})(F) - \Omega^{n-2}(u_3, \ldots, u_{n-i})(F) = 0$$

Q.E.D.

Remark 1.4. The set $E(n,i+1)$ of Proposition 1.2 can be identified with the total space of an $(i+1)$-plane bundle over the Stiefel variety $St^n_{n-i-1}(k)$ by using the inner product on k^n. In a similar way, the set $F(n,i,j)$ can be identified with the total space of the bundle of unit vectors in a line bundle over the Stiefel variety $St^n_{n-i-2}(k)$. More specifically, each $(u;W;e(W))$ in $F(n,i,j)$ can be parametrized by $(u_1, \ldots, u_{j-1}, x, y, u_{j+2}, \ldots, u_{n-i}) \in St^n_{n-i}(k)$. $x \wedge y = e(W)$ is a unit vector in the 1-dimensional k-vector space $\Lambda^2_k(W)$. Usually, these bundles are equipped with some topology. In the present setting, it appears to be natural to consider them with the discrete topology. In this manner, the appearance of the Eilenberg-MacLane cohomology of $SO(n;k)$ is to be expected.

2. COINDUCED MODULES.

For ease of reference, we recall some wellknown facts about coinduced modules, compare Sah [59; II, section 5].

Let H be any subgroup of an abstract group G. Let M be any left H-module. The left G-module $\mathrm{coind}_H^G M$ is defined by:

$$\mathrm{coind}_H^G M = \{f \in \mathrm{Map}(G, M) \mid f(xy) = y^{-1} f(x), \ x \in G\}$$

The left G-action is defined by:

$$(xf)(y) = f(x^{-1}y), \ x, \ y \in G.$$

We note that Map(G, M) admits the involution sending f onto $f^{\#}$ with $f^{\#}(x) = f(x^{-1})$. The coinduced module can also be defined in terms of $f^{\#}$ (as is often done in the literature).

We will describe an equivalent definition. Let H act freely from the right on the Cartesian product $G \times M$ via the formula:

$$(g, m)h = (gh, h^{-1}m), \ h \in H, \ (g, m) \in G \times M.$$

Let $G \times_H M$ denote the space of all H-orbits on $G \times M$. We have the first factor projection map:

$$\mathrm{pr} : \quad G \times_H M \longrightarrow G/H.$$

Evidently pr is surjective and the fibre over a point of G/H is canonically isomorphic to M through the second factor projection map. We can view pr as a bundle projection map with fibre M and base G/H (all topologies are discrete). Let $\Gamma(G \times_H M)$ denote the space of all (discrete) sections. The additive structure of M shows that $\Gamma(G \times_H M)$ is an abelian group. If we let G act on $G \times M$ through left multiplication on the first factor G, then pr is a G-map and $\Gamma(G \times_H M)$ becomes a left G-module through the definition:

$$(xs)(p) = x(s(x^{-1}p)), \ p \in G/H, \ s \in \Gamma(G \times_H M), \ x \in G.$$

<u>Proposition 2.1</u>. With the preceding notations, $\Gamma(G \times_H M)$ is naturally isomorphic to $\mathrm{coind}_H^G M$ as a left G-module.

Proof. Let $s \in \Gamma(G \times_H M)$. Define $f: G \longrightarrow M$ so that $(g, f(g)) \in s(gH)$, $g \in G$. If $h \in H$, then $(gh, f(gh))$ must be H-equivalent to $(g, f(g))$. Therefore:

$$f(gh) = h^{-1} f(g), \quad g \in G \text{ and } h \in H.$$

This shows that $f \in \mathrm{coind}_H^G M$. The argument can be reversed. Moreover, if $x \in G$, then we have:

$$(g, (xf)(g)) = (g, f(x^{-1}g)) = x(x^{-1}g, f(x^{-1}g)) \in x(s(x^{-1}gH)).$$

We therefore have a G-module isomorphism. Q. E. D.

Using the associativity formula involving the functor Hom, Shapiro's Lemma takes the following form:

Proposition 2.2. With the preceding notations:

$$H^t(G, \Gamma(G \times_H M)) \cong H^t(H, M), \quad t \geq 0.$$

The following elementary fact has been observed by many people and is sometimes referred to as "center kills". A slightly more general version can be found in Sah [57; Proposition 2.7].

Proposition 2.3. Let M be a left G-module. Suppose there is an element σ in the center of G so that $\sigma - \mathrm{Id}_M$ is an automorphism of M. Then:

$$H^t(G, M) = 0 \text{ holds for all } t \geq 0.$$

Combining Propositions 2.2, 2.3 with straightforward homological algebra, we have the following "transgression" result:

Proposition 2.4. Let k be a field. Let $H(i)$, $0 \leq i \leq t$, be subgroups of a group G. Suppose $M(i)$ is a $kH(i)$-module so that a suitable element of the center of $H(i)$ is represented by a nontrivial scalar multiplication on $M(i)$. Suppose further that we have an exact sequence of left kG-modules:

$$0 \longrightarrow N \longrightarrow \mathrm{coind}_{H(t)}^G M(t) \longrightarrow \cdots \longrightarrow \mathrm{coind}_{H(0)}^G M(0) \longrightarrow M \longrightarrow 0.$$

We then have:

$$H^s(G, N) = 0, \ 0 \leq s \leq t; \text{ and } H^{s+t+1}(G, N) \cong H^s(G, M), \ s \geq 0.$$

3. ZEROTH SYZYGY.

We begin the study of $J^n_i(k) = \mathrm{Hom}_k(\mathcal{G}^n_i, k)$, $1 \leq i \leq n$. The Hadwiger map identifies $\mathcal{G}^n_i$ with a k-subspace of the k-vector space of finitely supported k-valued functions on $St^n_{n-i}(k)$. By duality, $J^n_i(k)$ can be identified with a k-quotient space of the k-vector space of all k-valued functions on $St^n_{n-i}(k)$. If $f \in \mathrm{Map}(St^n_{n-i}(k), k)$, then we use multi-vector notation to define $c_0(f)$ by:

$$c_0(f) = \Sigma_u f(u)\Omega^n(u) = \Sigma_u f(u)\Sigma_\eta \mathrm{sgn}(\eta)\chi^n(\eta\cdot u), \text{ where}$$
$$u = (u_1, \ldots, u_{n-i}) \in St^n_{n-i}(k),\ \eta = (\eta_1, \ldots, \eta_{n-i}),\ \eta_s = \pm 1, \qquad (3.1)$$
$$\mathrm{sgn}(\eta) = \eta_1 \cdot \ldots \cdot \eta_{n-s} \text{ and } \eta\cdot u = (\eta_1 u_1, \ldots, \eta_{n-i}u_{n-i}).$$

We then have a surjective k-linear map:

$$c_0 : \mathrm{Map}(St^n_{n-i}(k), k) \longrightarrow J^n_i(k).$$

Both $\mathrm{Map}(St^n_{n-i}(k), k)$ and $J^n_i(k)$ are $kO(n;k)$-modules. If $\sigma \in O(n;k)$, then σ acts on $St^n_{n-i}(k)$ and $\mathcal{P}^n$ through the formulas:

$$\sigma(u_1, \ldots, u_{n-i}) = (\sigma\cdot u_1, \ldots, \sigma\cdot u_{n-i});$$
$$\sigma[/x_1/\ldots/x_n/] = [/\sigma\cdot x_1/\ldots/\sigma\cdot x_n/].$$

As usual, σ acts on functions through the formulas:

$$(\sigma f)(u) = f(\sigma^{-1}\cdot u),\ u \in St^n_{n-i}(k),\ f \in \mathrm{Map}(St^n_{n-i}(k), k);$$
$$(\sigma g)[A] = g(\sigma^{-1}[A]),\ [A] \in \mathcal{P}^n,\ g \in J^n_i(k).$$

The map c_0 is then a $kO(n;k)$-module homomorphism. Since k has characteristic 0, relation (R_0) together with averaging allow us to replace the module $\mathrm{Map}(St^n_{n-i}(k), k)$ by the submodule of all alternating maps. An element f in $\mathrm{Map}(St^n_{n-i}(k), k)$ is called <u>alternating</u> if it satisfies:

$$f(\eta\cdot u) = \mathrm{sgn}(\eta)f(u),\ u \in St^n_{n-i}(k),\ \eta = (\eta_1, \ldots, \eta_{n-i}),\ \eta_s = \pm 1.$$

Let $S(n-i, i;k)$ be the subgroup of $O(n;k)$ formed by all matrices with form:

$$\mathrm{diag}(\eta_1, \ldots, \eta_{n-i}, \rho),\ \eta_s = \pm 1,\ \rho \in O(i;k). \qquad (3.2)$$

Let $e_1, \ldots, e_n$ be the standard unit column vectors in k^n, then the $kO(n;k)$-submodule of all alternating maps from $St^n_{n-i}(k)$ to k can be identified with the coinduced module:

$$Q_0 = \operatorname{coind}_{S(n-i,i;k)}^{O(n;k)} k(e_1 \wedge \dots \wedge e_{n-i}) \tag{3.3}$$

If $S^+(n-i,i;k) = S(n-i,i;k) \cap SO(n;k)$, then we also have:

$$Q_0 = \operatorname{coind}_{S^+(n-i,i;k)}^{SO(n;k)} k(e_1 \wedge \dots \wedge e_{n-i}). \tag{3.4}$$

Proposition 3.5. Let k be an ordered, square root closed field. With the notations of (3.1) through (3.4), we have an exact sequence of $kO(n;k)$- or $kSO(n;k)$-modules:

$$Q_0 \longrightarrow J_i^n(k) \longrightarrow 0, \quad 1 \leq i \leq n. \tag{3.6}$$

If $i = n$, then c_0 is an isomorphism of 1-dimensional k-vector spaces with $O(n;k)$ acting trivially. If $i < n$, then:

$$\begin{aligned} &H^t(O(n;k), Q_0) = 0 \text{ for all } t; \\ &H^t(SO(n;k), Q_0) = 0 \text{ for all } t \text{ when } i \text{ is odd; and} \\ &H^t(SO(n;k), Q_0) \cong H^t(O(i;k), \det) \text{ for all } t \text{ when } i \\ &\quad \text{is even and } \det = \Lambda^i_k(k^i) \text{ as } O(i;k)\text{-module.} \end{aligned} \tag{3.7}$$

Proof. We only need to verify (3.7). The first two assertions of (3.7) are results of Propositions 2.2, 2.3 together with (3.3) and (3.4) respectively. For the last assertion of (3.7), we conclude from Proposition 2.2 and (3.4) that $H^t(SO(n;k), Q_0) \cong H^t(S^+(n-i,i;k), k(e_1 \wedge \dots \wedge e_{n-i}))$, $t \geq 0$. $S^+(n-i,i;k)$ is the direct product of $C = \{\operatorname{diag}(\eta_1, \dots, \eta_{n-i}, I_i) \mid \eta_s = \pm 1 \text{ and } \operatorname{sgn}(\eta) = 1\}$ and $O(i;k) \cong \{\operatorname{diag}(\det \rho, 1, \dots, 1, \rho) \mid \rho \in O(i;k)\}$. Since C acts trivially on $k(e_1 \wedge \dots \wedge e_{n-i})$ and k has characteristic 0, an averaging process leads to the isomorphism: $H^t(S^+(n-i,i;k), k(e_1 \wedge \dots \wedge e_{n-i})) \cong H^t(O(i;k), \det)$, $t \geq 0$. An alternative argument can be given by using the Hochschild-Serre spectral sequence associated to the exact sequence:

$$1 \longrightarrow C \longrightarrow S^+(n-i,i;k) \longrightarrow O(i;k) \longrightarrow 1$$

We omit the details. Q.E.D.

Remark 3.8. In general, $H^t(SO(i;k), k)$ is not always 0 for $t > 1$, see the works of Cheeger [18], Harris [36], Sah-Wagoner [61] among others. As a result, we do not know the general structure of $H^t(O(i;k), \det)$.

4. DUAL BASES FOR $\mathcal{G}_i^n$ AND $J_i^n(k)$.

The results in the present section are implicitly contained in sections 4.2 and 4.3. We rephrase them by using the square root closure of k. For a general vector space of infinite dimension over a field k, a vector space basis is usually constructed by invoking Zorn's Lemma. If B is such a basis, then a "dual" basis B* is formed by characteristic functions of one element subsets of B. It is a dual basis in the sense that every element of the dual space can be viewed as an element of Map(B, k) and be identified with an unrestricted sum of elements from B* with coefficients from k. In the book by Hadwiger [31], $\mathcal{G}_i^n$ is viewed as a $\mathbb{Q}$-vector space. As a result, a $\mathbb{Q}$-vector space basis for $\mathcal{G}_i^n$ can only be found in the abstract sense just described. With the k-vector space structure on $\mathcal{G}_i^n$ given through the Hadwiger map, we can determine k-basis for $\mathcal{G}_i^n$ in a geometric manner. It involves some choices.

Let u_0 be an arbitrarily prescribed unit vector in k^n. Let B_1^n be the set of all classes in $\mathcal{G}_1^n$ represented by orthogonal n-simplices of the form:

$$/x_1/\dots/x_n/,\ x_1 = -\|x_1\|\, u_0,\ \|x_1+\dots+x_n\| = 1 \qquad (4.1)$$

Such an n-simplex will be called a <u>unit</u> <u>u_0-negative</u> <u>orthogonal</u> <u>n-simplex</u>. Let B_1^{*n} be the set of all Hadwiger invariants $\Omega^n(u_1, \dots, u_{n-1})$ so that:

$$u_0, u_1, \dots, u_{n-1} \text{ are k-independent in } k^n,\ \langle u_0, u_i\rangle \in k^+ \text{ for } 1 \le i \le n-1. \qquad (4.2)$$

The (n-1)-frame $(u_1, \dots, u_{n-1})$ with property (4.2) will be called <u>u_0-positive</u>. It determines a unique n-frame $(u_1, \dots, u_n)$ with $\langle u_0, u_i\rangle \in\,]0,1[$, $1 \le i \le n$. Conversely, each such n-frame leads to n u_0-positive (n-1)-frames through the omission of u_i, $1 \le i \le n$. We can restate Lemma 4.2.1.

<u>Proposition 4.3</u>. Let k be an ordered, square root closed field. Let u_0 be any unit vector in k^n. The sets B_1^n and B_1^{*n} described by (4.1) and (4.2) form dual bases for the k-vector spaces $\mathcal{G}_1^n$ and $J_1^n(k)$.

A k-basis for $\mathcal{G}_i^n$, $i > 1$, can now be constructed through induction. We note that B_1^n depends on the choice of a unit vector u_0 in k^n. For each subspace

U containing u_0, we have a basis $B^m_1(U)$ for $\mathcal{G}^m_1(U)$, $m = \dim_k U \geq 1$. Let W be the orthogonal complement of U in k^n. $\mathcal{G}^{n-m}_{i-1}(W)$ is nonzero if and only if $1 \leq i-1 \leq n-m$. By induction, we assume that a k-basis for $\mathcal{G}^{n-m}_{i-1}(W)$ has been constructed. Let B^n_i denote the classes of orthogonal basic i-fold cylinders of the following form:

$$A_1 \# \ldots \# A_i,\ \text{Lsp}\, A_1 = U, \text{ the class of } A_1 \text{ lies in } B^m_1(U); \text{ the class of } A_2 \# \ldots \# A_i \text{ lies in } B^{n-m}_{i-1}(W), W \text{ is the orthogonal complement of } U \text{ in } k^n; \text{ and } U \text{ ranges over k-subspaces of dimension } m \text{ so that } u_0 \in U \text{ and } 1 \leq i-1 \leq n-m. \tag{4.4}$$

The following result can be extracted from section 4.3.

<u>Proposition 4.5</u>. Let k be an ordered, square root closed field. Let u_0 be a unit vector in k^n. Let $1 < i \leq n$ and let U ranges over k-subspaces of k^n with the properties:

$$u_0 \in U,\ i-1 \leq n-m,\ m = \dim_k U. \tag{4.6}$$

Then $\mathcal{G}^n_i$ is the direct sum of k-subspaces $\mathcal{G}^n_{i,U}$ with $\mathcal{G}^n_{i,U}$ isomorphic to $\mathcal{G}^m_1(U) \otimes_k \mathcal{G}^{n-m}_{i-1}(W)$ through the Minkowski sum map and W denotes the orthogonal complement of U in k^n. In particular, the inductively defined set B^n_i is a k-basis for $\mathcal{G}^n_i$.

It is easy to see that B^n_n consists of the class of the unit n-cube. We will now define a dual basis $B*^n_i$, $1 \leq i \leq n$, for $J^n_i(k)$. Except when i = 1 or n, $B*^n_i$ will usually not be dual to B^n_i. The definition is again inductive. For any subspace U of k^n containing u_0, $B*^m_1(U)$, $m = \dim_k U$, is defined in accordance with (4.2). If W denote the orthogonal complement of U, then $B*^{n-m}_{i-1}(W)$ is assumed to be a set of Hadwiger invariants defined through induction. Let $1 < i-1 \leq n$. Define $B*^n_i$ to consist of all $\Omega^n(u_1, \ldots, u_{n-i})$ so:

$$\text{For (unique) } s \text{ with } 1 \leq s \leq n-i+1, \text{ we have } \langle u_0, u_j \rangle \in k^+, 1 \leq j \leq s-1;\ \langle u_0, u_j \rangle = 0,\ s \leq j \leq n-i;\ \dim_k U = s, \text{ where } U = \text{Lsp}(u_0, u_1, \ldots, u_{s-1}); \text{ and } \Omega^{n-s}(u_s, \ldots, u_{n-i}) \text{ lies in } B*^{n-s}_{i-1}(W), \text{ where } W \text{ is the orthogonal complement of } U \text{ in } k^n. \tag{4.7}$$

Proposition 4.8. Let k be an ordered, square root closed field. Let u_0 be a unit vector in k^n. Let $B*^n_i$, $1 \leq i \leq n$, be defined in accordance with (4.2) and (4.6). Then the k-basis B^n_i for $\mathcal{G}^n_i$ defined in Proposition 4.5 can be modified through a Gram-Schmidt procedure to yield a new k-basis B'^n_i for $\mathcal{G}^n_i$ so that $B*^n_i$ is the dual basis to B'^n_i for $J^n_i(k)$.

The proof is contained in section 4.3. The main point is simply the fact that at most a finite number of Hadwiger invariants can be nonzero on any given polyhedron of dimension n in k^n.

5. FIRST SYZYGY. WEIGHT $i = n-1$.

We begin the study of ker c_0 with c_0 defined in section 3. As noted in Proposition 3.5, c_0 is an isomorphism when the weight i is n. We consider the next case when the weight i is n-1. In this case, the relations (R_2) are vacuous.

Proposition 5.1. Let k be an ordered, square root closed field. With the notations of section 3, we have the following exact sequence of kO(n;k)-modules:

$$0 \longrightarrow k^n \xrightarrow{\iota} Q_0 \xrightarrow{c_0} J^n_{n-1}(k) \longrightarrow 0, \quad 1 < n.$$

For $w \in k^n$, $\iota(w)$: $St^n_1(k) \longrightarrow k$ is the alternating map defined by:

$$\iota(w)(u) = < w, u >, \quad u \in St^n_1(k).$$

Proof. Relation (R_1) shows that $c_0 \circ \iota = 0$. The nondegeneracy of $< , >$ shows that ι is injective. If $\rho \in O(n;k)$, then:

$$\rho(\iota(w))(u) = \iota(w)(\rho^{-1} \cdot u) = < w, \rho^{-1} \cdot u > = < \rho \cdot w, u >.$$

It remains for us to show that ker $c_0 \subset$ im ι. Let $f \in$ ker c_0 so that:

$$f : St^n_1(k) \longrightarrow k, \ f(-u) = -f(u) \text{ and } \Sigma_u f(u)\, \Omega^n(u) = 0$$
where $u \in St^n_1(k)$.

We select $x \in St^n_1(k)$ arbitrarily and set $g = f - f(x)\,\iota(x)$. It follows that $g \in$ ker c_0 and $g(x) = 0$. We therefore have:

$$0 = \Sigma_u g(u)\Omega^n(u) = 2\, \Sigma_{<u,x> \in\, [0,1[}\, g(u)\,\Omega^n(u).$$

Let W be the orthogonal complement of kx and let B be an arbitrary (n-1)-simplex in W. The evaluation of $\Sigma_u g(u)\Omega^n(u)$ on /x/# B yields:

$$\Sigma_{u\in W}\, g(u)\,\Omega^{n-1}(u)(B) = 0.$$

By induction on n, we can find $y \in W$ so that:

$$g(u) = \langle y, u \rangle,\ u \in St_1^{n-1}(W).$$

If $h = g - \iota(y)$, then $h \in \ker c_0$ and we have:

$$0 = \Sigma_u h(u)\,\Omega^n(u) = 2\,\Sigma_{\langle u,x\rangle \in]0,1[}\, h(u)\,\Omega^n(u),$$

$$= \Sigma_U \Sigma_{u\in U,\ \langle u,x\rangle\in]0,1[}\, h(u)\,\Omega^n(u),\ \text{where}$$

$\|u\| = 1$ and U ranges over all 2-dimensional subspaces with $x \in U$.

Let $u_0 = x$ in Propositions 4.3 and 4.5. B^{*n}_{n-1} is then the set of all $\Omega^n(u)$ with $\langle u, x\rangle \in]0,1[$. The preceding equation shows that h = 0 so that we have $f = \iota(f(x)x + y)$. Q.E.D.

Proposition 5.2. Let k be an ordered, square root closed field. Then:

(a) $H^t(O(n;k), k^n) = H^t(O(n;k), J^n_{n-1}(k)) = 0$ for all $t \geq 0$;

(b) If n is even, then $H^t(SO(n;k), k^n) = H^t(SO(n;k), J^n_{n-1}(k)) = 0,\ t \geq 0$;

(c) If n is odd and $(t,n) \neq (1,3)$, then $H^t(SO(n;k), k^n) = 0,\ t = 0, 1$;

(d) $H^1(O(2;k), \det) \cong H^1(SO(2,k), k) \cong \mathrm{Hom}(SO(2,k), k)$.

Proof. (a) and (b) follow from Proposition 2.3 because $-I_n$ acts as -1 on k^n. (c) is clear when n = 1 or when t = 0. In general, $\mathcal{G}^n_i(E^+(n;k)) = \mathcal{G}^n_i(E(n;k))$ as shown in Theorem 1.4.1. Taking k-dual, we have:

$$H^0(SO(n;k), J^n_i(k)) = H^0(O(n;k), J^n_i(k)) \tag{5.3}$$

From (a) and the long cohomology sequence, we have the exact sequence:

$$0 \longrightarrow H^1(SO(n;k), k^n) \longrightarrow H^1(SO(n;k), Q_0) \longrightarrow H^1(SO(n;k), J^n_{n-1}(k)).$$

Let n = 2m+1 be odd. We deduce from (3.7):

$$H^1(SO(2m+1;k), Q_0) \cong H^1(O(2m;k), \det) \tag{5.4}$$

The right hand side of (5.4) can be determined by using the Hochschild-Serre spectral sequence associated to the split exact sequence of groups:

$$1 \longrightarrow SO(s;k) \longrightarrow O(s;k) \longrightarrow \{\pm 1\} \longrightarrow 1$$

Since k is a field of characteristic 0, we have:

$$H^1(O(s;k), \det) \cong H^0(\{\pm 1\}, H^1(SO(s;k), \det)) \tag{5.5}$$

Since SO(s;k) acts trivially on det $\cong$ k, we have:

$$H^1(SO(s;k), \det) \cong \mathrm{Hom}(SO(s;k), k) \tag{5.6}$$

When $s \geq 3$, the Cartan-Dieudonne theorem, see Artin [5; Theorem 3.22], shows that SO(s;k) is generated by elements of order 2 corresponding to 180° rotations. Since k has characteristic 0, we have:

$$\mathrm{Hom}(SO(s;k), k) = 0 \text{ when } s \neq 2$$

Feeding this back into (5.4) through (5.6), we get (c). When s = 2, any element in O(2;k) - SO(2;k) acts as -Id on SO(2;k) and det. (d) therefore follows from (5.5) and (5.6). Q.E.D.

6. FIRST SYZYGY. WEIGHT $i < n-1$.

We continue with the preceding notations and consider ker c_0 when the weight i is less than n-1. It follows that $n \geq 3$ and relations (R_2) come into play. We first consider relations (R_1). Let $Q_{1,n-i}$ be the kO(n;k)-module:

$$Q_{1,n-i} = \mathrm{coind}_{S(n-i-1,i+1;k)}^{O(n;k)}(k(e_1 \wedge \ldots \wedge e_{n-i-1}) \otimes_k (ke_{n-i} + \ldots + ke_n))$$

As in the case with Q_0, we have the kSO(n;k)-module isomorphism:

$$Q_{1,n-i} \cong \mathrm{coind}_{S^+(n-i-1,i+1;k)}^{SO(n;k)}(k(e_1 \wedge \ldots \wedge e_{n-i-1}) \otimes_k (ke_{n-i} + \ldots + ke_n))$$

Moreover, $Q_{1,n-i}$ can be identified with the space of all maps:

$f: St^n_{n-i-1}(k) \longrightarrow k^n$ so that:

$f(u)$ is orthogonal to $Lsp(u)$ and $f(\eta \cdot u) = \mathrm{sgn}\, \eta \cdot f(u)$ (6.1)

where $u = (u_1, \ldots, u_{n-i}) \in St^n_{n-i-1}(k)$

For each f satisfying (6.1), define a map:

$$c_{1,n-i}(f) : St^n_{n-i}(k) \longrightarrow k \tag{6.2}$$

so that:

$$c_{1,n-i}(f)(u,w) = \langle f(u), w \rangle, \text{ where} \\ (u,w) = (u_1, \ldots, u_{n-i-1}, w) \in St^n_{n-i}(k). \tag{6.3}$$

Proposition 6.4. Let k be an ordered, square root closed field. With the preceding notations, we have an exact sequence of $kO(n;k)$-modules:

$$0 \longrightarrow Q_{1,n-i} \xrightarrow{c_{1,n-i}} Q_0$$

Moreover, $c_0 \circ c_{1,n-i} = 0$ and we have:

(a) $H^t(O(n;k), Q_{1,n-i}) = 0$ for all $t \geq 0$;

(b) If $i+1$ is even, then $H^t(SO(n;k), Q_{1,n-i}) = 0$ for all $t \geq 0$;

(c) If $i+1$ is odd and $(t, i+1) \neq (1,3)$, then $H^t(SO(n;k), Q_{1,n-i}) = 0$, $t = 0, 1$;

(d) If $i+1 = 3$, then $H^1(SO(n;k), Q_{1,n-i}) \cong H^1(SO(3;k), k^3)$.

Proof. According to (6.3), $c_{1,n-i}(f)$ is alternating, therefore lies in Q_0. According to (6.1), f(u) is orthogonal to Lsp(u) and w freely ranges over the unit vectors in the orthogonal complement of Lsp(u), the injectivity of $c_{1,n-i}$ therefore follows from the nondegeneracy of $\langle\ ,\ \rangle$. If $\sigma \in O(n;k)$, then:

$$c_{1,n-i}(\sigma f)(u,w) = \langle (\sigma f)(u), w \rangle = \langle \sigma(f(\sigma^{-1} \cdot u)), w \rangle \\ = \langle f(\sigma^{-1} \cdot u), \sigma^{-1} \cdot w \rangle = (\sigma(c_{1,n-i}(f)))(u,w).$$

$c_{1,n-i}$ is therefore a $kO(n;k)$-module injection. From relations (R_1), we obtain:

$$c_0 \circ c_{1,n-i}(f) = \Sigma_{(u,w)} \langle f(u), w \rangle \Omega^n(u,w) = \Sigma_u \Sigma_w \langle f(u), w \rangle \Omega^n(u,w) = 0.$$

Assertions (a) and (b) are easy consequences of Propositions 2.2 and 2.3. More precisely, $\mathrm{diag}(-1, I_{n-1})$ and $\mathrm{diag}(I_{n-i-1}, -I_{i+1})$ lie in the center of $S(n-i-1, i+1;k)$. The latter lies in $S^+(n-i-1, i+1;k)$ when $i+1$ is even. These act according to $-\mathrm{Id}$ on $k(e_1 \wedge \ldots \wedge e_{n-i-1}) \otimes_k (ke_{n-i-1} + \ldots + ke_n)$. In cases (c) and (d), a spectral sequence argument together with Proposition 2.2 shows:

$$H^t(SO(n;k), Q_{1,n-i}) \cong H^t(O(i+1;k), \det \otimes_k k^{i+1}) \\ \cong H^0(\{\pm 1\}, H^1(SO(i+1;k), \det \otimes_k k^{i+1}))$$

(c) and (d) follow as in Proposition 5.2. Q.E.D.

Relations (R_2) depend on the choice of an integer j with $1 \leq j \leq n-i-1$. If j is such an integer, then let $T(j, n-i, i;k)$ be the subgroup of $O(n;k)$ formed by all matrices of the form:

$$\mathrm{diag}(\eta_1, \ldots, \eta_{j-1}, \rho, \eta_{j+2}, \ldots, \eta_{n-i}, \sigma), \text{ where } \eta_s = \pm 1,$$
$$\rho \in O(2;k), \ \sigma \in O(i;k), \text{ and } 1 \leq j \leq n-i-1$$

Let $Q_{1,j}$ be the $kO(n;k)$-module:

$$Q_{1,j} = \mathrm{coind}_{T(j,n-i,i;k)}^{O(n;k)}(k(e_1 \wedge \ldots \wedge e_{n-i})), \ 1 \leq j \leq n-i-1.$$

As before, we have the $kSO(n;k)$-module isomorphism:

$$Q_{1,j} \cong \mathrm{coind}_{T^+(j,n-i,i;k)}^{SO(n;k)}(k(e_1 \wedge \ldots \wedge e_{n-i})), \ 1 \leq j \leq n-i-1.$$

Similarly, $Q_{1,j}$ can be identified with the space of all maps:

$f : St_{n-i}^n(k) \longrightarrow k$ so that:

$$f(u \cdot \mathrm{diag}(\eta_1, \ldots, \eta_{j-1}, \rho, \eta_{j+2}, \ldots, \eta_{n-i})) = \mathrm{sgn}\,\eta \cdot \det\rho \cdot f(u) \qquad (6.5)$$
where $u = (u_1, \ldots, u_{n-i}) \in St_{n-i}^n(k)$.

It is immediate that $Q_{1,j}$ is a $kO(n;k)$-submodule of Q_0. We let $c_{1,j}$ denote the inclusion map. For $1 \leq j \leq n-i-1$, the subgroups $T(j, n-i, i;k)$ are conjugate in $O(n;k)$. It follows that $Q_{1,j}$, $1 \leq j \leq n-i-1$, are pairwise isomorphic as $kO(n;k)$- or $kSO(n;k)$-modules.

Proposition 6.6. Let k be an ordered, square root closed field. With the preceding notations, $c_0 \circ c_{1,j} = 0$, $1 \leq j \leq n-i-1$.

Proof. Let f satisfy (6.5). Rewrite $c_0(f)$ as a triple sum:

$$c_0(f) = \Sigma_{(x,y)} \Sigma_W \Sigma_{u \wedge v = e(W)} f(x, u, v, y) \Omega^n(x, u, v, y)$$

where $(x, y) = (x_1, \ldots, x_{j-1}, y_{j+2}, \ldots, y_{n-i})$ ranges over $St_{n-i-2}^n(k)$, W ranges over oriented 2-planes in the orthogonal complement of $Lsp(x, y)$ with $e(W)$ denoting one of the two possible unit vectors in $\Lambda_k^2(W)$, and (u, v) ranges over all 2-frames in W with the same orientation as $e(W)$. (6.5) shows that $f(x, u, v, y)$ depends only on (x, y). Relation (R_2) shows that the innermost sum is 0. Q.E.D.

From now on, we let Q_1 be the direct sum of the kO(n;k)-modules $Q_{1,j}$, $1 \leq j \leq n-i$. We define $c_1 : Q_1 \longrightarrow Q_0$ so that it is just $c_{1,j}$ on $Q_{1,j}$. We therefore have the following sequence of kO(n;k)-modules and homomorphisms:

$$Q_1 \xrightarrow{c_1} Q_0 \xrightarrow{c_0} J_i^n(k) \longrightarrow 0, \; 1 \leq i \leq n-2, \; c_0 \circ c_1 = 0$$

Proposition 6.7. Let k be an ordered, square root closed field. With the preceding notations, assume i = n-2. Then $\ker c_1 \cong \Lambda_k^2(k^n)$.

Proof. Suppose i = n-2. Q_1 is then the direct sum of $Q_{1,1}$ and $Q_{1,2}$ and $c_{1,1}$, $c_{1,2}$ are known to be injections. It follows that $\ker c_1$ is isomorphic to $Q_{1,1} \cap Q_{1,2}$. The latter is just the space of all maps:

$$\begin{aligned} &f: St_1^n(k) \longrightarrow k^n \text{ so that: (a) } f(-u) = -f(u); \\ &\text{(b) } \langle u, f(u) \rangle = 0; \text{ and (c) } \langle f(u), v \rangle = \det\rho \cdot \langle f(x), y \rangle \qquad (6.8) \\ &\text{where } \rho \in O(2;k) \text{ so that } (u, v) = (x, y)\cdot\rho \in St_2^n(k). \end{aligned}$$

Let w, $z \in k^n$. Define $f(w, z) : St_1^n(k) \longrightarrow k^n$ by:

$$f(w, z)(u) = \langle w, u \rangle z - \langle z, u \rangle w \qquad (6.9)$$

It is immediate that f(w, z) satisfies (a) and (b) of (6.8). In general, the inner product < , > on k^n induces an inner product on $\Lambda_k^j(k^n)$, $1 \leq j \leq n$, by:

$$\langle x_1 \wedge \ldots \wedge x_j, y_1 \wedge \ldots \wedge y_j \rangle = \det(\langle x_p, y_q \rangle),$$
$$\text{where } 1 \leq p, q \leq j \text{ and } x_p, y_q \in k^n$$

When j = 2, $\langle f(w, z)(u), v \rangle = \langle w \wedge z, u \wedge v \rangle$. It follows that (c) of (6.8) also holds for f(w, z). (6.9) also shows that f(w, z) is k-bilinear in z, w and that f(w, w) = 0 for all w in k^n. As a consequence, the map sending (z, w) onto f(z, w) induced a k-linear map:

$$\iota : \Lambda_k^2(k^n) \longrightarrow Q_{1,1} \cap Q_{1,2} \cong \ker c_1 \qquad (6.10)$$

We assert that ι is a kO(n;k)-isomorphism.

Let $\sigma \in O(n;k)$. We then have:

$$\begin{aligned} (\sigma(f(w, z)))(u) &= \sigma\{f(w, z)(\sigma^{-1} \cdot u)\} = \langle w, \sigma^{-1} \cdot u \rangle \sigma \cdot z - \langle z, \sigma^{-1} \cdot u \rangle \sigma \cdot w \\ &= \langle \sigma \cdot w, u \rangle \sigma \cdot z - \langle \sigma \cdot z, u \rangle \sigma \cdot w = f(\sigma \cdot w, \sigma \cdot z)(u). \end{aligned}$$

It follows that ι is kO(n;k)-module homomorphism. Since $n \geq 2$ and $<\ ,\ >$ is nondegenerate, ι is not the zero map. It is wellknown that $\Lambda_k^2(k^n)$ is an absolutely irreducible kO(n;k)-module. It follows that ι is an injective map. We still need to show the surjectivity of ι. Let f satisfy (6. 8). Since k is square root closed, each ray of k^n contains a unique unit vector. We can therefore extend f to a map (also denoted by f):

$$f: k^n \longrightarrow k^n \text{ so that } f(0) = 0 \text{ and}$$
$$f(\lambda u) = \lambda f(u),\ \lambda \in k^+,\ u \in St_1^n(k). \tag{6. 11}$$

From (a) and (b) of (6. 8), the extended map f has the properties:

$$\text{(a)}\ \ f(\alpha x) = \alpha f(x),\ \alpha \in k,\ x \in k^n; \text{ and}$$
$$\text{(b)}\ \ <x, f(x)> = 0,\ x \in k^n \tag{6. 12}$$

For $x,\ y \in k^n$, consider $g(x, y) \in k$ given by:

$$g(x, y) = < f(x), y > \tag{6. 13}$$

It is immediate that g is k-homogeneous in x and k-linear in y. Suppose that x, y are nonzero vectors in k^n with x_1, y_1 denoting the unit vectors on k^+x and k^+y respectively. From (b) of (6. 12), we have:

$$g(x, y) = g(x, y - <y, x_1> x_1) \tag{6. 14}$$

In particular, $g(x, y) = 0$ when x and y are k-dependent. If x and y are k-independent, then we can combine the k-homogeneity with (6. 14) and (c) of (6. 8) to obtain:

$$g(x, y) = \epsilon \, \| x \wedge y \| \, g(u, v), \text{ where } (u, v) \text{ is any 2-frame}$$
$$\text{with } x \wedge y = \epsilon \, \| x \wedge y \| \, u \wedge v,\ \epsilon = \pm 1.$$

It follows that:

$$g(x, y) = -g(y, x),\ x,\ y \in k^n \text{ and } g(x, y) = 0 \text{ if } x \wedge y = 0. \tag{6. 15}$$

Consequently, g(x, y) is k-bilinear in x, y and alternating. The extended map f in (6. 11) is therefore k-linear. (6. 13) and (6. 15) imply that f is skew symmetric in the sense that:

$$< f(x), y > = - < x, f(y) >, \quad x, y \in k^n.$$

The nondegenerate inner product $< , >$ allows us to identify $\mathrm{Hom}_k(k^n, k^n)$ with $k^n \otimes_k k^n$ as kSO(n;k)-module. Under this identification, the space of all skew symmetric maps is identifiable with $\Lambda_k^2(k^n)$. The surjectivity of ι is then a consequence of the injectivity of ι and dimension counting. Q.E.D.

Theorem 6.16. Let k be an ordered, square root closed field. With the preceding notations, let $1 \leq i \leq n-2$. We then have an exact sequence of kO(n;k)-modules:

$$Q_1 \xrightarrow{c_1} Q_0 \xrightarrow{c_0} J^n_{n-i}(k) \longrightarrow 0$$

When $i = n-2$, $\ker c_1 \cong \Lambda_k^2(k^n)$.

In view of the preceding results, we only need to prove $\ker c_0 \subset \mathrm{im}\, c_1$. For this purpose, we stratify $St^n_{n-i}(k)$. This stratification depends on the choice of a unit vector u_0 in k^n. We fix such a choice. If $u = (u_1, \ldots, u_{n-i})$ is in $St^n_{n-i}(k)$, then u_0 can be written uniquely in the form:

$$u_0 = w + \Sigma_j < u_0, u_j > u_j, \quad 1 \leq j \leq n-i, \text{ w orthogonal to } Lsp(u) \qquad (6.17)$$

Let $S(t, \epsilon)$ be the subset of all u in $St^n_{n-i}(k)$ such that:

$$\begin{aligned} &< u_0, u_j > \neq 0 \text{ for exactly t indices j, } 1 \leq j \leq n-i, \text{ and} \\ &\epsilon = 0, 1 \text{ according to w in (6.17) is 0 or not.} \end{aligned} \qquad (6.18)$$

We call (t, ϵ) the order of u and we order (t, ϵ) lexicographically. We let $S_0(t, \epsilon)$ denote the subset of $St^n_{n-i}(k)$ formed by those $u \in St^n_{n-i}(k)$ such that:

$$\mathrm{ord}(u) = (t, \epsilon) \text{ and } < u_0, u_j > \neq 0 \text{ for } 1 \leq j \leq t. \qquad (6.19)$$

We note that $S_0(0, 0) = \emptyset$ and $S_0(n-i, 1) = S(n-i, 1)$. In order to keep track of our arguments, we define ord(f) for $f \in Q_0$ by:

$$\mathrm{ord}(f) = \min\{\mathrm{ord}(u) \mid u \in \mathrm{supp}(f)\}, \quad \mathrm{ord}(0) = \infty.$$

For any subset C of Q_0, we define ord(C) by:

$$\mathrm{ord}(C) = \max\{\mathrm{ord}(f) \mid f \in C\}.$$

We note that $\mathrm{im}\, c_1$ is the unique coset C in $Q_0/\mathrm{im}\, c_1$ with $\mathrm{ord}(C) = \infty$.

We next introduce the Weyl group generators $r_j \in O(n-i;k)$, $1 \leq j \leq n-i-1$. These are defined by their actions on $St^n_{n-i}(k)$ from the right:

$$(u_1, \ldots, u_{n-i})r_j = (u_1, \ldots, u_{j-1}, -u_{j+1}, u_j, u_{j+2}, \ldots, u_{n-i}).$$

Together with all diagonal matrices $\eta = \mathrm{diag}\{\eta_1, \ldots, \eta_{n-i}\}$, they generate the Weyl group of type C_{n-i}. Modulo the diagonal subgroup, the r_j's generate the full permutation group Σ_{n-i}. In a similar way, $T(j, n-i, i;k)$ acts on the right of $St^n_{n-i}(k)$ with kernel $O(i;k)$. We can rephrase relations (R_2):

Let $u = (u_1, \ldots, u_{n-i}) \in St^n_{n-i}(k)$, $1 \leq i \leq n-2$. Let $\mathrm{ord}(u) = (t, \epsilon)$, $1 \leq t \leq n-i$, $\epsilon \in \{0, 1\}$ with respect to $u_0 \in St^n_1(k)$. Let $1 \leq j \leq n-i-1$. Then there exists a unique g in $Q_{1,j}$ with: (6.20)

(a) $g(u) = 1$, $\mathrm{supp}(g) = T(j, n-i, i;k)$-orbit of u; and

(b) $\mathrm{supp}(g) \subset S(t+1, \epsilon) \cup \eta$-orbit of u and $u \cdot r_j$.

To see this, we use the alternating character of the Hadwiger invariants to write (R_2) in the equivalent form:

$$\Omega^n(\ldots, u_j, u_{j+1}, \ldots) + \Omega^n(\ldots, -u_{j+1}, u_j, \ldots)$$

$$= - \Sigma_{(x, y)}\Omega^n(\ldots, x, y, \ldots)$$

where (x, y) ranges over all 2-frames in $\mathrm{Lsp}(u_j, u_{j+1})$ with $\langle u_j, x\rangle \in]0, 1[$ and $x \wedge y = u_j \wedge u_{j+1}$. In particular, $|\langle u_j, x\rangle|$, $|\langle u_j, y\rangle|$, $|\langle u_{j+1}, x\rangle|$, $|\langle u_{j+1}, y\rangle| \in]0, 1[$.

Each such relation then pulls back to a unique g in $Q_{1,j}$ by $c_{1,j}$. We can also rephrase relations (R_1):

Let $u = (u_1, \ldots, u_{n-i}) \in St^n_{n-i}(k)$, $1 \leq i \leq n-2$. Let $\mathrm{ord}(u) = (t, 0)$, $1 \leq t \leq n-i$, with respect to $u_0 \in St^n_1(k)$. Assume $\langle u_{n-i}, u_0 \rangle \neq 0$. Then there exists g in $\mathrm{im}\, c_{1,n-i}$ with:

$g(u) = 1$, $\mathrm{supp}(g) \subset S(t, 1) \cup \eta$-orbit of u. (6.21)

In fact, g can be taken to be the image of $f \in Q_{1,n-i}$, where $f : St^n_{n-i-1}(k) \longrightarrow k^n$, $f(u_1, \ldots, u_{n-i-1}) = u_{n-i}$ and $\mathrm{supp}(f)$ is the η-orbit of $(u_1, \ldots, u_{n-i-1})$.

To see this, we again use the alternating character of Hadwiger invariants to rewrite (R_1) in the equivalent form:

$$\Omega^n(u_1, \ldots, u_{n-i}) = -\Sigma_x <u_{n-i}, x> \Omega^n(u_1, \ldots, u_{n-i-1}, x)$$

where x ranges over all unit vectors orthogonal to $Lsp(u_1, \ldots, u_{n-i-1})$ and such that $<u_{n-i}, x> \in]0, 1[$.

This relation corresponds to $g = c_{1,n-i}(f)$ as indicated in (6.21). The conditions $ord(u) = (t, 0)$ and $<u_0, u_{n-i}> \neq 0$ translate to:

$$u_0 = \Sigma_{1 \leq j \leq n-i} <u_0, u_j> u_j, \quad <u_0, u_{n-i}> \neq 0.$$

If x is any unit vector in the orthogonal complement of $Lsp(u_1, \ldots, u_{n-i-1})$ with $<u_{n-i}, x> \in]0, 1[$, then $kx + ku_{n-i}$ is a 2-dimensional subspace in the orthogonal complement of $Lsp(u_1, \ldots, u_{n-i-1})$. If x and y is a 2-frame in $kx + ku_{n-i}$, then $<u_{n-i}, x> \in]0, 1[$ shows that:

$$<u_0, u_{n-i}> u_{n-i} = <u_0, u_{n-i}> (<u_{n-i}, x> x + <u_{n-i}, y> y)$$

where $<u_{n-i}, x> \neq 0 \neq <u_{n-i}, y>$.

It follows that:

$ord(u_1, \ldots, u_{n-i-1}, x) = (t, 1)$ when x is any unit vector orthogonal to $Lsp(u_1, \ldots, u_{n-i-1})$ with $<u_{n-i}, x> \in]0, 1[$.

This yields the assertion on the support of g in (6.21).

Proposition 6.22. Let k be an ordered, square root closed field. Let u_0 be an arbitrary unit vector in k^n. Assume $1 \leq i \leq n-2$. With the preceding notations, let C be any nonzero coset in $Q_0/\text{im } c_1$. Then:

(a) $ord(C) = (t, 1)$ for some $0 \leq t \leq n-i$; and

(b) there exists f in C with $ord(f) = ord(C)$ and $supp(f) \cap S(t, 1) \subset S_0(t, 1)$.

Proof. Since $C \neq \text{im } c_1$, $ord(C) = (t, \epsilon)$ is not ∞. Since $S(0, 0) = \phi$, (t, ϵ) is not (0, 0). With only a finite number of possibilities for orders, we can find f in C so that $ord(f) = ord(C)$. We will now modify f by elements of $\text{im } c_1$.

Suppose that $u = (u_1, \ldots, u_{n-i}) \in \text{supp}(f) \cap S(t, \epsilon)$ and satisfies:

$$\text{exactly one of } u_j,\ u_{j+1} \text{ is orthogonal to } u_0. \tag{6.23}$$

In particular, $t > 0$. If $g \in Q_{1,j}$ is defined in accordance with (6.20), $f' = f - f(u)\cdot c_{1,j}(g) \in C$ has the same order as f, the set $\text{supp}(f') \cap S(t, \epsilon)$ and the set $\text{supp}(f) \cap S(t, \epsilon)$ are identical except for the fact the η-orbit of u and the η-orbit of $u\cdot r_j$ are exchanged. Repeated applications imply that we can assume:

$$\text{supp}(f) \cap S(t, \epsilon) \subset S_0(t, \epsilon) \text{ whenever } t > 0. \tag{6.24}$$

Here we must be careful because we may have to apply the modification an infinite number of times. We need to make sure that there is no infinite stacking up on $S(t+1, \epsilon)$. Each $(\ldots, x, y, \ldots) \in S(t+1, \epsilon)$ appearing in (b) of (6.20) may arise in at most a finite number of ways from $u \in S(t, \epsilon)$. More precisely, there are only a finite number of possibilities for j. Once this is fixed, x, y must be a 2-frame in $\text{Lsp}(u_j, u_{j+1})$ and (u_j, u_{j+1}) must satisfy (6.23). Since we have a 2-dimensional space, there are only 8 possibilities for (u_j, u_{j+1}), these are $(\underline{+}u_j, \underline{+}u_{j+1})$ and $(\underline{+}u_{j+1}, \underline{+}u_j)$. As a consequence, there is no infinite stacking up on $S(t+1, \epsilon)$. The last observation also shows that there is no infinite stacking up on $S(t, \epsilon)$. Since r_j's generate Σ_{n-i} mod the diagonal subgroup formed by the η's, we can reach (6.24) with repeated applications of modifications by elements of im c_1.

We will now show $\epsilon = 1$. Since $S(0, 0) = \phi$, we may assume $t > 0$. The argument of the preceding paragraph can be applied to $u = (u_1, \ldots, u_{n-i})$ in $\text{supp}(f) \cap S(t, 0)$ to yield the condition $< u_0, u_{n-i} > \neq 0$. If $g \in Q_{1,n-i}$ is taken in accordance with (6.21), $f' = f - f(u)\cdot c_{1,n-i}(g) \in C$ has the order (t, ϵ'). If $\text{ord}(C) = (t, 0)$, then $\epsilon' = 0$ and $\text{supp}(f') \cap S(t, 0) = \text{supp}(f) \cap S(t, 0) - \{u\}$. If we repeat this argument for each point of $\text{supp}(f) \cap S(t, 0)$, we will ultimately reach a contradiction. However, we must again take care so that there is no infinite stacking up. For a fixed choice of $(u_1, \ldots, u_{n-i-1})$, the number of unit vectors u_{n-i} with $(u_1, \ldots, u_{n-i}) \in S(t, 0)$ and with $< u_0, u_{n-i} > \neq 0$ is 2. To be precise, u_{n-i} must be a unit vector on $k(u_0 - \Sigma_{1 \leq j \leq n-i-1} < u_0, u_j > u_j)$.

As a consequence, there is no infinite stacking up. The eventual contradiction shows that ϵ must be 1 when $\text{ord}(f) = \text{ord}(C) = (t, \epsilon)$. Q. E. D.

Proof of Theorem 6.16. As indicated, we only need to show $\ker c_0 \subset \text{im } c_1$. Suppose false. Let $f \in \ker c_0 - \text{im } c_1$ so that:

(a) $\text{ord}(f) = \text{ord}(f + \text{im } c_1) = \text{ord}(\ker c_0 - \text{im } c_1) = (t, 1)$, $0 \leq t \leq n - i$; and (6.25)

(b) $\text{supp}(f) \cap S(t, 1) = \text{supp}(f) \cap S_0(t, 1) \neq \phi$

The existence of f follows from Proposition 6.22. Ultimately, we will get a contradiction to either (a) or (b) of (6.25). We recall that $1 \leq i \leq n-2$. Our argument will proceed by induction on i.

Suppose $i = 1$. Let $u = (u_1, \ldots, u_{n-1}) \in \text{supp}(f) \cap S_0(t, 1)$. There are two possibilities:

Case 1. $t = n-1$. $S_0(t, 1) = S(n-1, 1)$ and the η-orbits on $S(n-1, 1)$ are in bijective correspondence with B^{*n}_1 described in (4.2). According to Proposition 4.3, $f \in \ker c_0$ and $\text{supp}(f) \subset S(n-1, 1)$ if and only if $f = 0$. $t = n-1$ together with (a) of (6.25) force $\text{supp}(f) \subset S(n-1, 1)$. We therefore have a contradiction to (b) of (6.25).

Case 2. $t < n-1$. From (b) of (6.25), we have $\langle u_0, u_{n-1} \rangle = 0$. Let $(u_1, \ldots, u_n)$ be one of the two n-frames extending u. ku_{n-1} is then the unique 1-dimensional subspace of $\text{Lsp}(u_{n-1}, u_n)$ orthogonal to u_0. This is a consequence of $\epsilon = 1$ so that $\langle u_0, u_n \rangle \neq 0$. As in (6.21), we again rephrase (R_1):

Let $u = (u_1, \ldots, u_{n-i}) \in St^n_{n-i}(k)$, $1 \leq i \leq n-2$. Let $\text{ord}(u) = (t, 1)$, $0 \leq t \leq n-i-1$, with respect to $u_0 \in St^n_1(k)$. Assume $\langle u_0, u_{n-i} \rangle = 0$. Then there exists g in $\text{im } c_{1, n-1}$ with:

$$g(u) = 1, \ \text{supp}(g) \subset S(t+1, 1) \cup \eta\text{-orbit of } u. \tag{6.26}$$

In fact, g can be taken to be the image of $f \in Q_{1, n-i}$, where $f : St^n_{n-i-1}(k) \longrightarrow k^n$, $f(u_1, \ldots, u_{n-i-1}) = u_{n-i}$ and $\text{supp}(f)$ is the η-orbit of $(u_1, \ldots, u_{n-i-1})$.

In the present case, $Lsp(u_{n-1}, u_n)$ is the orthogonal complement of the subspace $Lsp(u_1, \dots, u_{n-2})$ and ku_{n-1} is the unique line in this plane orthogonal to u_0. We can therefore repeat our modification of f by subtracting off $f(u) \cdot c_{1,n-i} g$ with g determined by u as indicated in (6.26). There is no infinite stacking up. We eventually find h in $f + \text{im } c_1$ with:

$$ord(h) > (t, 1) \text{ in the lexicographic ordering.}$$

This contradicts (a) of (6.25).

Suppose that $1 < i \leq n-2$ and that Theorem 6.16 has been proved for i-1 and all n. We first decompose f into a finite sum:

$$f = \Sigma_{(j,\epsilon)} f(j,\epsilon), \quad supp(f(j,\epsilon)) \subset S(j,\epsilon)$$

Applying Proposition 6.22 to each $f(j,\epsilon) + \text{im } c_1$, we may assume:

$$f = \Sigma_j f(j,1), \quad supp(f(j,1)) \subset S_0(j,1), \text{ and}$$
$$f(j,1) = 0 \text{ for } j < t \text{ and } f(t,1) \neq 0.$$

We again have two cases.

Case 1. $t = 0$. Let W denote the orthogonal complement of ku_0 so that $W \cong k^{n-1}$ and Theorem 6.16 has been proven for W with i-1 in place of i. Let $/u_0/\#B$ be an orthogonal basic i-fold cylinder so that B is just an arbitrary orthogonal basic (i-1)-fold cylinder in W. As summarized in Lemma 4.3.1, we have:

$$0 = c_0(f)[/u_0/\#B] = c_0(f(0,1))[/u_0/\#B] \tag{6.27}$$

Moreover, for each $(u_1, \dots, u_{n-i}) \in S(0,1)$, we have:

$$\Omega^n(u_1, \dots, u_{n-i})[/u_0/\#B] = \Omega^{n-1}(u_1, \dots, u_{n-i})[B]. \tag{6.28}$$

Since $S(0,1) \subset St^{n-1}_{n-i}(W)$, we can view f(0,1) as an element of $Q_0(W)$ with the obvious interpretations. (6.27) and (6.28) together imply that f(0,1) lies in the kernel of c_0 associated to $Q_0(W)$. Since $St^{n-1}_{n-i}(W) = St^{n-1}_{n-1-(i-1)}(W)$, we are in the situation where n and i are replaced by n-1 and i-1. We can now apply induction to f(0,1). This means that f(0,1) is in the image of c_1 associated to $Q_1(W)$. More precisely, f(0,1) is an infinite sum of terms built up from k-multiples of functions constructed according to relations (R_1) and (R_2) associated with W. A relation of type (R_2) associated with W becomes a relation of type (R_2) associated with k^n through the simple device of replacing Ω^{n-1} by Ω^n. In terms of the associated function $g \in Q_{1,j}(W)$, we extend $g: St^{n-1}_{n-i}(W) \longrightarrow k$ to $St^n_{n-i}(k)$ so that it is 0 outside of $St^{n-1}_{n-i}(W)$. If $(u_1, \ldots, u_{n-i}) \in St^{n-1}_{n-i}(W)$, then a relation of type (R_1) in W and in k^n has the form:

$$\Omega^{n-1}(u_1, \ldots, u_{n-i}) = - \Sigma_{w \in W} \langle u_{n-i}, w \rangle \, \Omega^{n-1}(u_1, \ldots, u_{n-i-1}, w);$$

$$\Omega^{n}(u_1, \ldots, u_{n-i}) = - \Sigma_{w \in W} \langle u_{n-i}, w \rangle \, \Omega^{n}(u_1, \ldots, u_{n-i-1}, w)$$

$$- \Sigma_{w \notin W} \langle u_{n-i}, w \rangle \, \Omega^{n}(u_1, \ldots, u_{n-i-1}, w);$$

where w ranges over unit vectors orthogonal to $u_1, \ldots, u_{n-i-1}$ and satisfies $\langle u_{n-i}, w \rangle \in \,]0,1[$.

It follows that the associated functions g in $Q_{1,n-i}(W)$ can be extended to a function from $St^n_{n-i-1}(k)$ to k^n so that it is 0 outside of $St^{n-1}_{n-i-1}(W)$. When $(u_1, \ldots, u_{n-i}) \in \mathrm{supp}(f(0,1))$, the preceding formulas show that we can find h in im c_1 so that:

$$\mathrm{supp}(f(0,1)-h) \subset S(1,0) \cup S(1,1)$$

This means that:

$$\mathrm{ord}(f-h) > (0,1) = \mathrm{ord}(f)$$

This contradicts (a) of (6.25).

Case 2. $t > 0$. Let $(u_1, \ldots, u_{n-i}) \in \mathrm{supp}(f(t,1)) \subset S_0(t,1)$. Since f is alternating, we can multiply u_j by ± 1 and assume:

$$\begin{aligned} &\langle u_0, u_j \rangle \in \,]0,1[\,, \; 1 \le j \le t; \\ &\langle u_0, u_j \rangle = 0, \; t+1 \le j \le n-i; \text{ and} \\ &u_0, u_1, \ldots, u_t \text{ are k-independent.} \end{aligned} \tag{6.29}$$

We note that $\epsilon = 1$ is needed to ensure the first and the third assertion of (6.29). As in Lemma 4.2.1, we can find a unit u_0-negative $(t+1)$-simplex $/x_1/\ldots/x_{t+1}/$ in $Lsp(u_0, u_1, \ldots, u_t)$ dual to $\Omega^{t+1}(u_1, \ldots, u_t)$. Consider an arbitrary orthogonal basic i-fold cylinder of the form $/x_1/\dots/x_{t+1}/\#B$ so that B is just an orthogonal basic (i-1)-fold cylinder in the orthogonal complement W of $\mathrm{Lsp}(u_0, u_1, \dots, u_t)$. As before, we have:

$$0 = c_0(f)[/x_1/\dots/x_{t+1}/\#B] = c_0(f(t,1))[/x_1/\dots/x_{t+1}/\#B] \quad (6.30)$$

Let $(v_1, \dots, v_{n-i}) \in S_0(t, 1)$. Then:

$$\begin{aligned} &\Omega^n(v_1, \dots, v_{n-i})[/x_1/\dots/x_{t+1}/\#B] = 0 \text{ when} \\ &\quad (v_1, \dots, v_t) \text{ is not in the } \eta \text{ orbit of } (u_1, \dots, u_t); \\ &\Omega^n(v_1, \dots, v_{n-i})[/x_1/\dots/x_{t+1}/\#B] = \qquad (6.31) \\ &\quad \eta_1 \cdot \dots \cdot \eta_t \cdot \Omega^{n-t-1}(v_{t+1}, \dots, v_{n-i})[B] \text{ when} \\ &\quad v_j = \eta_j u_j, \ 1 \le j \le t, \ \eta_j = \pm 1. \end{aligned}$$

We further partition $S_0(t, 1)$ into a disjoint union of subsets $S(u_1, \dots, u_t)$ with $(u_1, \dots, u_t)$ satisfying the first and third condition of (6.29). More precisely, $(v_1, \dots, v_{n-i}) \in S_0(t, 1)$ is placed in $S(u_1, \dots, u_t)$ if and only if:

$$v_j = \pm u_j, \ 1 \le j \le t.$$

We next decompose $f(t, 1)$ into the sum (possibly infinite):

$$\begin{aligned} &f(t,1) = \Sigma\, g(u_1, \dots, u_t), \text{ where } (u_1, \dots, u_t) \text{ satisfies} \\ &(6.29) \text{ and } \mathrm{supp}(g(u_1, \dots, u_t)) \subset S(u_1, \dots, u_t). \end{aligned}$$

By choice, at least one $g(u_1, \dots, u_t)$ is nonzero. We note that:

$$\begin{aligned} &2^{-t} c_0(g(u_1, \dots, u_t)) = \Sigma_v f(u_1, \dots, u_t, v)\Omega^n(u_1, \dots, u_t, v) \\ &\text{where } v = (v_{t+1}, \dots, v_{n-i}) \text{ ranges over } St^{n-t-1}_{n-i-t}(W). \end{aligned}$$

From (6.30) and (6.31), we have:

$$\begin{aligned} &\Sigma_v f(u_1, \dots, u_t, v)\Omega^{n-t-1}(v)[B] = 0, \ v \in St^{n-t-1}_{n-i-t}(W) \text{ and} \\ &B \text{ is an arbitrary orthogonal basic (i-1)-fold cylinder} \\ &\text{in } W. \end{aligned}$$

With $(u_1, \dots, u_t)$ fixed, $f(u_1, \dots, u_t, v)$ can be viewed as an alternating function on $St^{n-t-1}_{n-t-1-(i-1)}(W)$. We may now proceed by induction and follow the argument used in the preceding case. In exactly the same manner, we reach a contradiction to (a) of (6.25). Q.E.D.

7. SOME COHOMOLOGY CALCULATIONS.

We continue with the preceding notations.

Proposition 7.1. Let k be an ordered, square root closed field. Let $1 \leq i \leq n-2$ and $1 \leq j \leq n-i-1$.

(a) If $i < n-2$, then $H^t(O(n;k), Q_{1,j}) = 0$ for all $t \geq 0$;

(b) If $i = n-2$, then $H^t(O(n;k), Q_{1,j}) \cong \coprod_s H^s(O(2;k), \det) \otimes_k H^{t-s}(O(n-2;k), k)$, where $0 \leq s \leq t$;

(c) $H^{2m}(O(2;k), \det) = 0$ and $H^{2m+1}(O(2;k), \det) \cong H^{2m+1}(SO(2;k), k)$.

(d) If i is odd and $i < n-2$, then $H^t(SO(n;k), Q_{1,j}) = 0$ for all $t \geq 0$;

(e) If i is odd and $i = n-2$, then $H^t(SO(n;k), Q_{1,j})$
$\cong \coprod_s H^s(O(2;k), \det) \otimes_k H^{t-s}(SO(n-2;k), k)$, where $0 \leq s \leq t$;

(f) If i is even, then $H^t(SO(n;k), Q_{1,j})$
$\cong H^0(\{\pm 1\}, H^t(SO(2;k) \times SO(i;k), \det))$, where -1 acts by -1 on det and according to $O(2;k)/SO(2;k)$ on $SO(2;k)$ as well as $O(i;k)/SO(i;k)$ on $SO(i;k)$.

Proof. The vanishing assertions (a) and (d) follow from Propositions 2.2 and 2.3. In general, we conclude from Proposition 2.2 that:

$$H^t(O(n;k), Q_{1,j}) \cong H^t(T(j, n-i, i;k), k(e_1 \wedge \ldots \wedge e_{n-i})), \quad t \geq 0;$$
$$H^t(SO(n;k), Q_{1,j}) \cong H^t(T^+(j, n-i, i;k), k(e_1 \wedge \ldots \wedge e_{n-i})), \quad t \geq 0.$$

When $i = n-2$, $T(j, n-i, i;k) = S(2, n-2;k) \cong O(2;k) \times O(n-2;k)$. Assertion (b) follows from (7.2) and a spectral sequence argument. More precisely, we need to know that the Hochschild-Serre spectral sequence associated to the direct product $O(2;k) \times O(n-2;k)$ has E_2-terms equal to E_∞-terms. Since k is a field of characteristic 0, all the higher differentials are zero because they correspond to cup product with suitable torsion cohomology classes, see Sah [59] for detailed discussions. For assertions (e) and (f), the argument is similar. Using the assumption that k has characteristic 0, we can average over any finite subgroup of the center which acts trivially on the coefficient. The spectral sequence argument then takes over. For (c), we use spectral sequence to deduce: (see also [39].)

$$H^s(O(2;k), \det) \cong H^0(\{\pm 1\}, H^s(SO(2;k), \det)), \quad s \geq 0.$$

Again, we have used the fact that $H^t(G, V) = 0$ holds for $t > 0$ when G is a finite group and V is a vector space over a field of characteristic 0. Now $SO(2;k)$ acts trivially on $\det \cong k$ so that $H^s(SO(2;k), \det)$ is just the k-dual of $H_s(SO(2;k), \det)$. Since $SO(2;k)$ is a commutative group and $\det \cong k$ is a field of characteristic 0, $H_s(SO(2;k), \det)$ can be calculated through a direct limit argument over the finitely generated subgroups of $SO(2;k)$. We can even assume these subgroups to be free abelian because k is a field of characteristic 0. However, for free abelian groups, the s-dimensional homology is generated by the 1-dimensional classes through the Pontrjagin product. It is now evident that -1 acts according to $(-1)^{s+1}$ on $H_s(SO(2;k), \det)$. By duality, it also acts according to $(-1)^{s+1}$ on $H^s(SO(2;k), \det)$. (c) therefore follows. Q.E.D.

Exactly the same type of argument used to prove (c) of Proposition 7.1 also shows that:

$$\begin{aligned} &H^{2m}(O(2;k), k) \cong H^{2m}(SO(2;k), k) \text{ and} \\ &H^{2m+1}(O(2;k), k) = 0, \text{ where } m \geq 0. \end{aligned} \tag{7.3}$$

The structure of $H^s(SO(2;k);k)$ is as follows:

$$\begin{aligned} &H^0(SO(2;k);k) \cong k; \text{ and more generally,} \\ &H^s(SO(2;k);k) \text{ is the k-dual of } \Lambda_k^s(k \otimes SO(2;k)). \end{aligned} \tag{7.4}$$

Proposition 7.5. Let k be an ordered, square root closed field. Let $n > 2$. There is then an exact sequence:

$$\begin{aligned} 0 &\longrightarrow H^1(O(n;k), \Lambda_k^2(k^n)) \longrightarrow \mathrm{Hom}(SO(2;k), k) \\ &\longrightarrow H^0(O(n;k), J_{n-2}^n) \longrightarrow H^2(O(n;k), \Lambda_k^2(k^n)) \longrightarrow 0. \end{aligned}$$

If n is odd, then $O(n;k)$ may be replaced by $SO(n;k)$ in the exact sequence. If K is an ordered, square root closed extension field of k, then the above exact sequence is compatible with respect to restriction maps. In general, $H^0(O(n;k), J_{n-2}^n)$ is the k-dual of a k-vector space of dimension equal to the cardinality of k.

Proof. We let i = n-2 in Theorem 6.16. From the vanishing results of Propositions 3.5 and 6.4, the long cohomology sequence yields the exact sequence:

$$0 \longrightarrow H^1(O(n;k), \Lambda_k^2(k^n)) \longrightarrow H^1(O(n;k), Q_{1,1}) \longrightarrow H^0(O(n;k), J^n_{n-2}(k))$$
$$\longrightarrow H^2(O(n;k), \Lambda_k^2(k^n)) \longrightarrow H^2(O(n;k), Q_{1,1}) \longrightarrow H^1(O(n;k), J^n_{n-2}(k))$$
$$\longrightarrow \dots$$

From (b), (c) of Proposition 7.1 and (7.3), $H^1(O(n;k), Q_{1,1}) \cong \mathrm{Hom}(SO(2;k),k)$ and $H^2(O(n;k), Q_{1,1}) = 0$. The desired exact sequence then follows. When n is odd, O(n;k) is the direct product of SO(n;k) and $\{\pm I_n\}$. Since $-I_n$ acts trivially on $\Lambda_k^2(k^n)$ and $J^n_{n-2}(k)$, O(n;k) may be replaced by SO(n;k) throughout. The compatibility of restriction maps with respect to extension fields K of k follows easily from universal coefficient theorem. $H^0(O(n;k), J^n_{n-2}(k))$ is the k-dual of $\mathcal{G}^n_{n-2}(E(n;k))$. The dimension calculation in the Appendix shows that $\mathcal{G}^n_{n-2}(E(n;k))$ is a k-vector space of dimension equal to the cardinality of k.

Q.E.D.

Remark 7.6. When n = 3, $H^1(O(3;k), \Lambda_k^2(k^n)) \cong H^1(SO(3;k), k^3)$. In view of Proposition 5.2, (c), it seems reasonable to believe that $H^1(SO(3;k), k^3) = 0$. However, sporadic exceptions to general statements about low dimensional cohomology groups of finite Chevalley groups lend support to the belief that $H^1(SO(3;k), k^3)$ might in fact be nonzero. What is needed is the precise map from Hom(SO(2;k), k) to $H^0(O(3;k), J^3_1)$. In passing, we note that the results in Chapter 7 show that $H^0(O(n;k), J^n_{n-2})$ is stable for n > 2. Proposition 7.5 corrects some of the calculations made in our preliminary manuscript.

8. OPEN PROBLEMS.

It is apparent that we have left open two main problems. Both of them are somewhat technical in nature.

Higher Syzygy Problem. When $n \geq 4$, the kO(n;k)-module Q_1 in Theorem 6.16 is the direct sum of more than 2 coinduced modules. It is not clear what ker c_1 looks like. It would be quite pretty if the general pattern for weight i equals to n, n-1 and n-2 continues.

Eilenberg-MacLane Cohomology of Classical Groups. The calculations of the preceding section indicate that we need to have a systematic procedure to study the Eilenberg-MacLane cohomology groups of the form $H^i(G, M)$ where M ranges over wellknown representation spaces of the classical group G. The case that seems to be best understood is that of a finite group G of classical type. On the other hand, problems in geometry, number theory, etc. seem to indicate the presence of a large amount of mystery when G is a semisimple Lie group.

6 Spherical scissors congruence

In the present chapter, we lay a foundation for the scissors congruence problem in spherical spaces of arbitrary dimension. Historically, the spherical case arose after the Euclidean case. The origin can be traced to the area formula for a spherical triangle. Our foundation is quite similar to the point of view presented by Hadwiger [31] for the Euclidean spaces. A number of ideas contained in private communications from B. Jessen and A. Thorup have been incorporated. One of the principal motivating points traces back to Dehn: Whenever there is an invariant metric associated to the given scissors congruence data, an invariant can be formed by combining tangential data with normal data. Typically, tangential data and normal data are respectively given by volumes and angles in various dimensions. The success in the affine case can be attributed to the fact that both volumes and angles (in the guise of frames which involve only the concept of orthogonality) are algebraic. In terms of the present foundation, we single out a number of problems having some connections with the scissors congruence problem in spherical spaces of (topological) dimension greater than 2. As for the scissors congruence problem itself, we only have the classically known complete solutions in dimensions at most 2.

1. DEGREE GRADATION AND RING STRUCTURE.

Throughout this chapter, k will denote an Archimedean ordered field which is square root closed. With the Archimedean assumption, k can be identified with a subfield of the field $\mathbb{R}$ of real numbers. Some of the results are valid without the Archimedean hypothesis. When this is the case, we will either state it explicitly or place the word "Archimedean" in parentheses. k^n will denote the Euclidean n-space formed by column vectors of length n over k and will be equipped with the usual positive definite inner product $<,>$. We will embed k^n in k^{n+1} as the subspace of all vectors with last coordinate 0. The direct limit is then denoted by k^∞. Two similar models of the spherical space are available. The first is the familiar one in geometry and topology: $S(k^{n+1})$ is the n-sphere formed by all unit vectors in k^{n+1} with the (topological) dimension n and equipped with the group of motions $O(n+1;k)$. The direct limits are respectively $S(k^\infty)$ and $O(\infty;k)$ or simply $O(k)$. An n-simplex is then determined by any n+1 k-linearly independent unit vectors. If $x_0,\ldots,x_n$ is such a collection, then the associated spherical n-simplex is denoted by $\mathrm{sccl}\{x_0,\ldots,x_n\}$ with:

$$\mathrm{sccl}\{x_0,\ldots,x_n\} = \{x \in S(k^\infty) \mid x = \Sigma_i \alpha_i x_i,\ \alpha_i \geq 0,\ \alpha_i \in k\}.$$

For the second model, each $u \in S(k^\infty)$ is identified with the ray k^+u. As a result, a spherical n-simplex is identified with a polyhedral cone with apex (deleted) at the origin of k^∞. As such, it is the convex closure of n+1 k-linearly independent rays in k^∞. If we use closed rays rather than open rays, then the net change is the inclusion of the apex. In this latter description, it is also the intersection of a finite number of closed half spaces in a suitable n+1 dimensional k-subspace of k^∞. We will call the first the topological model and the second the ray model. Using the linear structure in the ray model, it is easy to see that we have scissors congruence data. For the purpose of defining a ring structure, it is more convenient to use the dimension arising from the ray model. Thus $\mathrm{sccl}\{x_0,\ldots,x_n\}$ is given the <u>degree</u> (or <u>ray dimension</u>) n+1. $\mathcal{P}(S(k^{n+1}), O(n+1;k))$ will be abbreviated to $\mathcal{P}S(k^{n+1})$ or $\mathcal{P}S^{n+1}$.

The inner product provides a k-valued invariant metric. When k is Archimedean, we can use the inverse cosine function to define an $\mathbb{R}$-valued invariant metric so that it is additive with respect to simple subdivision of 1-simplices. If $S(k^{n+1})$ is identified with the homogeneous space: $SO(n+1;k)/SO(n;k)$, then an invariant Haar measure yields an invariant $\mathbb{R}$-valued volume. Following L. Schläfli, the volume of $S(k^{n+1})$ is normalized to be 2^{n+1}. In particular, the distance between $x \in S(k^\infty)$ and its antipode $-x$ is normalized to be 2 (= volume of $S(k^1)$). With the Archimedean hypothesis on k, conditions (VP) and (VA) of Zylev's Theorem can be checked.

By using $S(k^{\infty})$, our scissors congruence data becomes infinite so that the elements of $\mathcal{P}S^{n+1}$ have a natural interpretation in this setting.

In the preceding paragraph, the Archimedean hypothesis played an essential role. Specifically, $SO(2;k)$ can be viewed as a quotient of a suitable subgroup of $\mathbb{R}$, + through the exponential map. In this manner, $\mathcal{P}S^2$ can be identified with a suitable subgroup of the group $\mathbb{R}$, +. When k is not Archimedean, we do not have an exponential map. Instead, we have a torsion free 2-divisible group $\mathcal{P}S^2$ with a nonsplit exact sequence:

$$0 \longrightarrow \mathbb{Z} \longrightarrow \mathcal{P}S^2 \longrightarrow SO(2;k) \longrightarrow 0 \qquad (1.1)$$

In (1.1), the generator 1 of $\mathbb{Z}$ is mapped onto the stable scissors congruence class of the unit circle $S(k^2)$. When $k = \mathbb{R}$, (1.1) becomes the wellknown exact sequence:

$$0 \longrightarrow \mathbb{Z} \longrightarrow \mathbb{R} \longrightarrow \mathbb{R}/\mathbb{Z} \longrightarrow 0$$

For the ring structure, we let ϕ be the (-1)-simplex. Its class plays the role of the multiplicative identity. The volume of ϕ is agreed to be 1. We therefore have $\mathcal{P}S^0 = \mathbb{Z}\cdot[\phi] \cong \mathbb{Z}$. We combine the spherical scissors congruence problem in all dimensions into a single problem by letting $\mathcal{P}S$ or $\mathcal{P}S(k^{\infty})$ be the direct sum of all $\mathcal{P}S^i$, $i \geq 0$. If $P = \mathrm{sccl}\{x_0, \ldots, x_p\}$ and $Q = \mathrm{sccl}\{y_0, \ldots, y_q\}$, then we find σ in $O(\infty;k)$ so that:

$$< Lsp(P), \sigma Lsp(Q) > = 0.$$

With such a σ at hand, we define the (orthogonal) <u>join</u> <u>product</u> * by the rule:

$$[P] * [Q] = [\,\mathrm{sccl}\{x_0, \ldots, x_p, \sigma y_0, \ldots, \sigma y_q\}].$$

It is then an easy matter to see that the * product is well defined and induces a commutative, $\mathbb{Z}$-graded (by degree) ring structure on $\mathcal{P}S$. The unit element is $[\phi]$. We also have the augmentation map sending $\mathcal{P}S$ onto $\mathbb{Z}$ with kernel $\mathcal{P}S^+ = \coprod_{i>0} \mathcal{P}S^i$.

In general, the geometric join $P*Q$ in $S(k^{\infty})$ can be defined as the spherical convex closure of $P \cup Q$ in $S(k^{\infty})$. When $Lsp(P) \cap Lsp(Q) = 0$, $P*Q$ is a spherical simplex whose vertices are the vertices of P and Q. If P, Q are identified with polyhedral cones in k^∞, then the cone associated to $P*Q$ is just the Minkowski sum of the corresponding cones.

Continuing with the Archimedean assumption on k, we assign to each p-simplex P in $S(k^\infty)$ its p-dimensional invariant volume $vol_p(P)$ in $\mathbb{R}$. To the class $[P] \in \mathcal{P}S^{p+1}$, we assign $vol_p(P)\cdot T^{p+1} \in \mathbb{R}[T]$. With our normalization of volumes, we have a ring homomorphism called the graded volume map:

$$\text{gr. vol.} : \mathcal{P}S(k^\infty) \longrightarrow \mathbb{R}[T]$$

Dehn's solution to the third problem of Hilbert extends to the spherical case: gr. vol. is not injective on $\mathcal{P}S(k^4)$. We will omit the exhibition of an example (the assertion is implicit in the Appendix).

2. FILTRATIONS.

We will consider three filtrations on $\mathcal{P}S$. The first is connected with the degree gradation. We simply set $\mathcal{P}S(i) = \Sigma_{0 \leq j \leq i} \mathcal{P}S^j$ and call it the degree filtration. The second is based on the close relationship between the orthogonal join and the orthogonal Minkowski sum and is suggested by B. Jessen and A. Thorup in private communications. The third is cruder than the second and originated from certain combinatorial aspects of the Gauss-Bonnet theorem. In contrast to the first, the second and third are decreasing filtrations rather than increasing.

Let $P = \text{sccl}\{x_0, \ldots, x_p\}$ be a spherical p-simplex in $S(k^\infty)$. P is called an i-fold (orthogonal) join when its vertices can be divided into i nonempty subsets which are mutually orthogonal in k^∞. Evidently, $1 \leq i \leq p+1 = \deg P$. We define $\mathcal{P}S_i^{p+1} = 0$ for $i > p+1$ and let $\mathcal{P}S_i^{p+1}$, $1 \leq i \leq p+1$, to be the subgroup of $\mathcal{P}S^{p+1}$ generated by the classes of the i-fold joins. Let $\mathcal{P}S_i = \coprod_{j \geq 0} \mathcal{P}S_i^j$. We then have:

$$\mathcal{P}S = \mathcal{P}S_1 \supset \mathcal{P}S_2 \supset \ldots, \quad \mathcal{P}_i^j * \mathcal{P}_s^t \subset \mathcal{P}_{i+s}^{j+t}, \; i, s \geq 1 \text{ and } j, t \geq 0.$$

The following proposition describes the additive structure of some of the groups in the join filtration.

Proposition 2.1. Let k be an Archimedean ordered, square root closed field. Then:

(a) $\mathcal{P}S^0 \cong \mathcal{P}S^1 \cong \mathbb{Z}$ and $\mathcal{P}S^i$ is 2-divisible for $i > 1$.

(b) If k is the square root closure of $\mathbb{Q}$ in $\mathbb{R}$, then $\mathcal{P}S^2$ is p-divisible for the prime p if and only if $p = 2$.

(c) The subring of $\mathcal{P}S$ generated by $\mathcal{P}S^1$ is isomorphic to $\mathbb{Z}[T]$ with the class of the unit n-sphere $S(k^{n+1})$ corresponding to $2^{n+1}T^{n+1}$, $n \geq -1$.

(d) If k is the algebraic closure of $\mathbb{Q}$ in $\mathbb{R}$, then the image of the graded volume map on the subring of $\mathcal{P}S$ generated by $\mathcal{P}S^2$ does not coincide with a polynomial ring in T^2 over a subring of $\mathbb{R}$ in positive degrees.

Proof. (a) $\mathcal{P}S^0$ and $\mathcal{P}S^1$ are free cyclic with generators $[\emptyset]$ and [point] by definition. The 2-divisibility of $\mathcal{P}S^i$ for $i > 1$ follows from Theorem 1.4.3. (b) follows from the classical translation of ruler and compass construction of angle divisions in the plane. (c) follows from the normalized graded volume map. For (d), let the image be $\Sigma_{i\geq 0} L_{2i} \cdot T^{2i}$ so that $L_0 = \mathbb{Z}$. For $i > 0$, L_{2i} is the additive subgroup of $\mathbb{R}$ generated by all i-fold products of elements of L_2. When $k = \mathbb{R}$, $L_{2i} = \mathbb{R}$ for $i > 0$. In our case, we have:

$$L_2 = \{\alpha \in \mathbb{R} \mid \exp(\pi\iota\alpha/2) \text{ is algebraic}\}.$$

(d) is equivalent with the statement that L_2 is not closed with respect to the product in $\mathbb{R}$. We first assert that L_2 is a $\mathbb{Q}$-subspace of $\mathbb{R}$ of countably infinite dimension over $\mathbb{Q}$. From its description, L_2 is a countable $\mathbb{Q}$-subspace of $\mathbb{R}$ containing $\mathbb{Q}$. $\mathbb{Q}$-linear independence in L_2 can be translated to multiplicative independence in $\mathbb{C}^\times$ through the exponential map. The ring $\mathbb{Z}[\iota]$ of Gaussian integers is a principal ideal domain such that a rational prime p in $\mathbb{Z}$ becomes the product of two nonassociate primes u_p and v_p in $\mathbb{Z}[\iota]$ if and only if $p \equiv 1 \bmod 4$. For these primes, u_p/v_p is an algebraic number of the form $\exp(\pi\iota\alpha_p/2)$, $\alpha_p \in \mathbb{R}$, because u_p/v_p has absolute value 1. The multiplicative independence of these u_p/v_p follows from the unique factorization theorem in $\mathbb{Z}[\iota]$. The corresponding α_p's are therefore $\mathbb{Q}$-linearly independent in L_2. The existence of an infinite number of rational primes congruent to 1 mod 4 then yields the countable infinitude of $\dim_{\mathbb{Q}} L_2$.

In order to show L_2 is not closed under product, we use the dimension calculation to find 1, α, β in L_2 which are $\mathbb{Q}$-linearly independent. It is enough to show that one of α^2, $\alpha\beta$, β^2 is not in L_2. By choice, we know that $\pi\iota/2$, $\pi\iota\alpha/2$, $\pi\iota\beta/2$ are also $\mathbb{Q}$-linearly independent. According to a theorem of Lang, see Baker [6 ; p. 119], one of $\exp(\pi\iota\gamma/2)$, γ lies in the set $\{\alpha, \beta, \alpha^2, \alpha\beta, \beta^2\}$, must be transcendental. Since $\alpha, \beta \in L_2$, this shows that one of α^2, $\alpha\beta$, β^2 is not in L_2. Q.E.D.

In the spherical case, the 0-simplex plays a distinguished role. In our normalization, $\mathrm{vol}_{p+1}(\{\text{point}\} * P) = \mathrm{vol}_p(P)$ holds for any p-simplex P in $S(k^\infty)$. We will call $\{\text{point}\} * P$ the (<u>orthogonal</u>) <u>spherical cone with base P</u>. As we will soon see, any 2m-dimensional spherical polyhedron is a linear combination (over $\mathbb{Z}$) of such cones in the sense of scissors congruence. The ideal in $\mathcal{P}S$ generated by [point] (equivalently, by $\mathcal{P}S^1$) will be denoted by $\mathcal{C}S$. As it stands, we have:

$$\mathcal{C}S = \coprod_{i>0} \mathcal{C}S^i, \quad \mathcal{C}S^i = [\text{point}] * \mathcal{P}S^{i-1}.$$

Closely related to the cone construction is the suspension construction. The orthogonal suspension is simply the orthogonal join with the 0-sphere. In view of the 2-divisibility of $\mathcal{P}S^i$, $i > 1$, $2\mathcal{C}S^j = \mathcal{C}S^j$ for $j > 2$ and $2\mathcal{C}S^j$ has index 2 in the infinite cyclic group $\mathcal{C}S^j$ for $j = 1, 2$. In terms of the solid spherical sector cut out by the associated polyhedral cone, an orthogonal suspension (of a 1-simplex) is often called a <u>lune</u> and a cone is just a <u>semi-lune</u>. For $i > 0$, let $\mathcal{C}_iS$ be the i-th power of $\mathcal{C}S$ (N.B. $\mathcal{C}_iS$ is in general smaller than $\mathcal{C}S_i = \mathcal{C}S \cap \mathcal{P}S_i$). These $\mathcal{C}_iS$ then define a decreasing filtration of $\mathcal{P}S$ called the <u>point-adic filtration</u> (or the <u>cone-adic filtration</u> or even the <u>semi-lune-adic filtration</u>).

<u>Proposition 2.2</u>. Let k be an (Archimedean) ordered, square root closed field. Then $\mathcal{P}S^{2i+1} = [\text{point}] * \mathcal{P}S^{2i} = \mathcal{C}S^{2i+1}$, $i \geq 0$. In particular, the ring $\mathcal{P}S/\mathcal{C}S$ is evenly graded by degree (= 1 + dimension).

<u>Proof</u>. By agreement, $\{\text{point}\} * \phi = \{\text{point}\}$ so that the assertion is clear for $i = 0$. Let $A = \mathrm{sccl}\{x_0, \ldots, x_t\}$ and view the vertices as ordered set of

t+1 points in $S(k^{\infty})$. For $0 \leq p \leq t$, let $\tau_p A = \mathrm{sccl}\{x_0, \ldots, -x_p, \ldots, x_t\}$. We therefore have:

$$\tau_p \circ \tau_q = \tau_q \circ \tau_p,\ \tau_p^2 = \mathrm{Id},\ \tau_0 \circ \ldots \circ \tau_p(A) = -\mathrm{Id}(A) \text{ and } A \coprod \tau_p(A) \text{ is a lune} \tag{2.3}$$

As a consequence, $[A] + (-1)^t[-\mathrm{Id}(A)] \in [\text{point}] * \mathcal{P}S^t$ for $t > 0$. If $t = 2i > 0$, then $\mathcal{P}S^{2i+1} = 2\mathcal{P}S^{2i+1} \subset [\text{point}] * \mathcal{P}S^{2i} = \mathcal{C}S^{2i+1} \subset \mathcal{P}S^{2i+1}$. Q. E. D.

Proposition 2.2 is a crude form of the Gauss-Bonnet theorem for polyhedra in spherical spaces. A more precise formulation requires a little bit of notation. Let $A = \mathrm{sccl}\{x_0, \ldots, x_n\}$ be a spherical n-simplex. Let $0 \leq i(0) < \ldots < i(p) \leq n$ and let $\epsilon(i(j)) = \pm 1$ for $0 \leq j \leq p$. We then define $\mathcal{L}_A(\epsilon(i(0))x_{i(0)}, \ldots, \epsilon(i(p))x_{i(p)})$ to be the set:

$$\{x \in S(k^{\infty}) \mid x = \Sigma_{0 \leq s \leq n} \alpha_s x_s,\ \alpha_s \in k,\ \epsilon(i(j))\alpha_{i(j)} \geq 0 \text{ for } 0 \leq j \leq p\}.$$

It is immediate that:

$\mathcal{L}_A(\epsilon(i(j))x_{i(j)})$ is a hemisphere in $S(\mathrm{Lsp}(A))$;

$\mathcal{L}_A(\epsilon(i(0))x_{i(0)}, \ldots, \epsilon(i(p))x_{i(p)}) = \bigcap_{0 \leq j \leq p} \mathcal{L}_A(\epsilon(i(j))x_{i(j)})$ is an (n-p)-fold orthogonal suspension of a suitable spherical p-simplex so that its class lies in $\mathcal{C}_{n-p}S \subset \mathcal{C}S$ when $p < n$.

$S(\mathrm{Lsp}(A)) = \coprod \mathcal{L}_A(\epsilon(0)x_0, \ldots, \epsilon(n)x_n)$, where $\epsilon(s)$ independently ranges over ± 1, $\mathcal{L}_A(\epsilon(0)x_0, \ldots, \epsilon(n)x_n) = \mathrm{sccl}\{\epsilon(0)x_0, \ldots, \epsilon(n)x_n\}$.

We can now apply the elementary inclusion-exclusion principle to $S(\mathrm{Lsp}(A))$ and the n+1 hemispheres $\mathcal{L}_A(x_i)$, $0 \leq i \leq n$. This then yields the following Gauss-Bonnet formula:

$$[\mathcal{L}_A(-x_0, \ldots, -x_n)] = [S(k^{n+1})] + \Sigma_p (-1)^{p+1} \Sigma_{(i)} [\mathcal{L}_A(x_{i(0)}, \ldots, x_{i(p)})] \tag{2.4}$$

where $0 \leq p \leq n$ and where (i) ranges over all sequences of integers: $0 \leq i(0) < \ldots < i(p) \leq n$.

According to n is even or odd, the unique term on the right hand side of (2.4) corresponding to p = n either reinforces or cancels the left hand side. The remaining terms on the right hand side are lunes; those corresponding to p = 0 are hemispheres and can be combined with $[S(k^{n+1})]$. We have:

Proposition 2.5. Let k be an ordered, square root closed field. Then $\mathcal{P}S^3 \cong \mathcal{P}S^2$. If k is not Archimedean, then there exist spherical polygons which are stably scissors congruent but not scissors congruent. If k is Archimedean, then the normalized area of the spherical triangle A is given by the real number:

(sum of the normalized interior angles at the vertices) - 2.

(Recall: the normalized volume of $S(k^{n+1})$ is 2^{n+1}.)

Proposition 2.6. Let k be an Archimedean ordered, square root closed field. Then:

(a) For $i \geq 2$, the subgroup $\mathcal{P}S^i_{i-1} = (C_{i-1}S)^i$ is isomorphic to $\mathcal{P}S^2$ through the graded volume map.

(b) When $k = \mathbb{R}$, $i \geq 2$, $\mathcal{P}S^i$ is the direct sum of $\mathcal{P}S^i_{i-1} \cong \mathbb{R}$ and the kernel of the graded volume map. The graded volume map is injective on $\mathcal{P}S^i$ for $i \leq 3$.

We omit the straightforward proofs. We note that $\mathcal{P}S^i_{i-1}$ is the analogue of $\mathcal{P}^{i-1}_{i-1}$ in the affine case and (b) of Proposition 2.6 uses completeness of $\mathbb{R}$.

3. SPHERICAL DUALITY, HOPF ALGEBRA AND DEHN INVARIANTS.

Let $A = \mathrm{sccl}\{x_0, \ldots, x_n\}$ be a spherical n-simplex in $S(k^\infty)$. We view the vertices $x_0, \ldots, x_n$ of A as an ordered set of n+1 k-linearly independent elements of $S(k^\infty)$. For each subset I of $\{0, \ldots, n\}$, A(I) will denote the face of A spanned by the vertices x_i with $i \in I$. By agreement $A(\emptyset) = \emptyset$. We will often identify A with the polyhedral cone in Lsp(A) spanned by A (with deleted apex at the origin of k^∞). Let y_i be the exterior unit normal in Lsp(A) to the codimensional 1 face opposite to x_i of the polyhedral cone associated to A. The ordered set of these exterior unit vectors $y_0, \ldots, y_n$

can be characterized by the conditions:

$$\begin{aligned}&\mathrm{Lsp}(y_0, \ldots, y_n) = \mathrm{Lsp}(x_0, \ldots, x_n) \cong k^{n+1},\\ &< y_i, y_i > = 1,\ < y_i, x_j > = 0 \text{ for } i \neq j, \text{ and } < x_i, y_i > < 0 \qquad (3.1)\\ &\quad \text{where } 0 \leq i,\ j \leq n.\end{aligned}$$

The spherical n-simplex $A^{\S} = \mathrm{sccl}\{y_0, \ldots, y_n\}$ is called the dual of A. If we identify A and $A^{\S}$ with closed polyhedral cones spanned by them in k^{∞}, then $A^{\S}$ can be identified as the set:

$$\{z \in \mathrm{Lsp}(A) \mid < z, x > \leq 0 \text{ for all } x \in A\}.$$

We clearly have:

$$\begin{aligned}&\sigma(A^{\S}) = (\sigma A)^{\S} \text{ for } \sigma \in O(\infty; k),\ (A^{\S})^{\S} = A,\ \phi^{\S} = \phi \text{ by agreement,}\\ &\{z\}^{\S} = \{-z\} \text{ for } z \in S(k^{\infty}), \text{ and } (A*B)^{\S} = A^{\S}*B^{\S} \text{ holds for any} \qquad (3.2)\\ &\text{orthogonal 2-fold join } A*B.\end{aligned}$$

Proposition 3.3. Let k be an (Archimedean) ordered, square root closed field. The map sending each spherical simplex A in $S(k^{\infty})$ onto its dual $A^{\S}$ induces a ring involution of $\mathcal{P}S/\mathcal{C}S$ preserving the degree gradation and the join filtration.

Proof. From (3.2), our map is orthogonal invariant, multiplicative and involutive. We need to know its behavior with respect to simple subdivision. Let $A = \mathrm{sccl}\{x_0, \ldots, x_n\}$, $n > 0$. Permuting the vertices when necessary, we can assume a simple subdivision of A is performed at $z = \alpha x_0 + \beta x_n \in S(k^{\infty})$ with $\alpha, \beta \in k^{+}$. This yields $\mathrm{sccl}\{x_0, \ldots, x_{n-1}, z\} = B$ and $\mathrm{sccl}\{z, x_1, \ldots, x_n\} = C$. Let $A^{\S} = \mathrm{sccl}\{y_0, \ldots, y_n\}$ as in (3.1). We then have:

$$\begin{aligned}&B^{\S} = \mathrm{sccl}\{w, y_1, \ldots, y_n\},\ C^{\S} = \mathrm{sccl}\{y_0, \ldots, y_{n-1}, -w\}\\ &\text{where } w = \gamma y_0 - \delta y_n \text{ with suitable } \gamma, \delta \in k^{+}.\end{aligned}$$

It is immediate that $\tau_n A = \mathrm{sccl}\{y_0, \ldots, y_{n-1}, -y_n\}$ is simply subdivided at w to yield $\tau_n B^{\S} = \mathrm{sccl}\{w, y_1, \ldots, y_{n-1}, -y_n\}$ and $\tau_n C^{\S} = \mathrm{sccl}\{y_0, \ldots, y_{n-1}, w\}$, where τ_n is defined in the proof of Proposition 2.2. Using (2.3), we get:

$$[A^{\S}] = [(B \coprod C)^{\S}] = [B^{\S}] + [C^{\S}] \bmod \mathcal{C}S.$$

The duality map permutes the generators of $\mathcal{P}S$ and stablizes the set of generators of $\mathcal{C}S$. (3.2) and the preceding equation imply that we have a well defined involution on $\mathcal{P}S/\mathcal{C}S$ preserving the degree gradation and the various filtrations. Q.E.D.

The involution defined in Proposition 3.3 will be called the <u>duality involution</u> on $\mathcal{P}S/\mathcal{C}S$. On spherical simplices A, we have:

$$[A]^{\S} = [A^{\S}]$$

This will be extended to spherical polyhedra through additivity and interior disjoint decomposition of the spherical polyhedra into simplices. We note that $\mathcal{P}S/\mathcal{C}S$ is evenly graded by degree (= 1 + dimension) so that the duality involution is of interest only on $\mathcal{P}S^{even} = \coprod_{i \geq 0} \mathcal{P}(S(k^{2i}), O(2i;k))$ mod $\mathcal{C}S$.

In the Gauss-Bonnet formula (2.4), $S(k^{n+1})$ was partitioned into 2^{n+1} n-simplices $sccl\{\epsilon(0)x_0, \ldots, \epsilon(n)x_n\}$, $\epsilon(i) = \pm 1$. During an oral conversation, Jeff Cheeger mentioned to us that this partition was used by him in proving a result of Whitney on characteristic classes, see Cheeger [17] as well as Halperin-Toledo [35]. We will now describe another partition. It can be traced back to L. Schläfli [63; vol. 1, p. 281]. The present proof is an adaptation of an observation by J. Milnor in a private communication concerning some of his results connected with classical integral geometry.

Let $A = sccl\{x_0, \ldots, x_n\}$ with $A^{\S} = sccl\{y_0, \ldots, y_n\}$ as in (3.1). For a subset I of $\{0, \ldots, n\}$ let I^c denote the complementary subset. Then:

<u>Proposition 3.4</u>. Let k be an (Archimedean) ordered, square root closed field. We have the following partition of the unit n-sphere in Lsp(A):

$$S(Lsp(A)) = \coprod_I A(I) * A^{\S}(I^c),$$ where I ranges over all 2^{n+1} distinct subsets of $\{0, \ldots, n\}$.

$A = A(\{0, \ldots, n\}) * A^{\S}(\emptyset)$, $A^{\S} = A(\emptyset) * A^{\S}(\{0, \ldots, n\})$, and $A(I) * A^{\S}(I^c)$ is an orthogonal 2-fold join when $I \neq \emptyset \neq I^c$.

<u>Proof</u>. We view A as a polyhedral cone in $Lsp(A) \cong k^{n+1}$. There is then a deformation retraction of Lsp(A) onto A sending each point of Lsp(A) along

a straight line segment towards the unique closest point in A. The location of the closest point is a standard problem in linear programming. In the present context, the procedure works for any ordered field. The square root closure is needed only in the identification of a ray with the unique unit vector on the ray. If we work backwards, the interior of the polyhedral cone corresponding to the simplex $A(I)*A^{\S}(I^c)$ is the full inverse image of the interior of the polyhedral cone corresponding to A(I) under the deformation retraction. Since A is the disjoint union of the interiors of the faces A(I), $I \subset \{0, \ldots, n\}$, we have the desired partition. The following diagram illustrates the partition and the deformation retraction for the case n = 1.

Q. E. D.

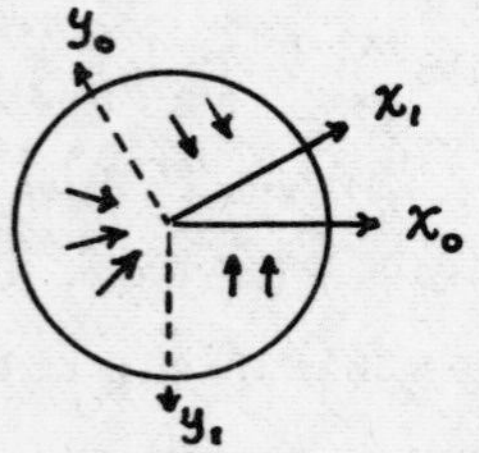

Suppose A is a spherical 1-simplex. It is evident from the preceding diagram that:

$$A \coprod A^{\S} \sim_{O(k)} \{\text{point}\} * S(k^1) = \text{half of a unit circle } S(k^2)$$

In general, let $A = \text{sccl}\{x_0, \ldots, x_n\}$ and let $A^{\S} = \text{sccl}\{y_0, \ldots, y_n\}$ as in (3.1) with $n > 0$. For $i \neq j$, $\text{sccl}\{y_i, y_j\}$ and the interior dihedral angle between the codimensional 1 faces of A opposite the vertices x_i and x_j add up to half of a unit circle $S(k^2)$ in the sense of scissors congruence. The next case of interest occurs when n = 3. The following proposition is due to Schläfli [63; vol. 1, p. 281].

<u>Proposition 3.5</u>. Let k be an Archimedean ordered, square root closed field. Let $A = \text{sccl}\{x_0, x_1, x_2, x_3\}$ with $A^{\S} = \text{sccl}\{y_0, y_1, y_2, y_3\}$. Then:

$$A \coprod A^{\S} \coprod \left(\coprod_{|I|=2} A(I)*A^{\S}(I^c) \right) \sim_{O(k)} \{\text{point}\} * S(k^3)$$
$$= \text{half of } S(k^4)$$

Proof. By Zylev's Theorem, stable scissors congruence is equivalent to scissors congruence. According to Proposition 3.4, the left hand side of our congruence is scissors congruent to lunes. Since n = 3, lunes lies in $\mathcal{P}S_3^4 = (\mathcal{C}_3 S)^4$ and the volume map in injective on $\mathcal{P}S_3^4$. It is therefore enough to verify the equality of volumes in our assertion. For this purpose, we view k as a subfield of $\mathbb{R}$ and calculate the differential of the volume of the left hand side. For a spherical n-simplex B, Schläfli's fundamental result [63; vol. 1, p. 235] states:

$$d(\mathrm{vol}(B)) = c_n \cdot \Sigma \, \mathrm{vol}_{n-2}(B^{(2)}) \, d\,\theta_B(B^{(2)}) \tag{3.6}$$

where c_n is a suitable constant depending on n and the normalizations of volumes, the summation ranges over all the codimensional 2 faces $B^{(2)}$ of B, and $\theta_B(B^{(2)})$ is the interior dihedral angle subtended at the codimensional 2 face $B^{(2)}$ of B. Using the assumption n = 3 together with the remark about $\mathrm{sccl}\{y_i, y_j\}$ preceding the proposition, the left hand side of our congruence has zero differential for its volume. If we continuously deform A until we reach $\mathrm{sccl}\{e_0, e_1, e_2, e_3\}$ with e_0, e_1, e_2, e_3 orthonormal, then it is easy to see that the total volume of the left hand side is equal to half of the total volume of $S(k^4)$. Q.E.D.

Up to now, $\mathcal{P}S/\mathcal{C}S$ is a commutative, augmented, evenly graded and filtered (by degree = 1 + dimension), $\mathbb{Z}$-algebra. In addition, it has a join filtration and a duality involution compatible with all the structures. We will now exhibit the existence of a Hopf algebra structure over $\mathbb{Z}$ with the duality involution playing the role of the antipode. This Hopf algebra structure is closely related to the Dehn invariant (to be studied in greater detail later).

Let $A = \mathrm{sccl}\{x_0, \ldots, x_n\}$ be a spherical n-simplex in $S(k^\infty)$. For each subset I of $\{0, \ldots, n\}$, let $\mathrm{Proj}_{A(I)}^{nor}$ denote the normal projection onto the orthogonal complement of $\mathrm{Lsp}(A(I))$. In the context of spherical geometry, this map induces a map (also denoted by $\mathrm{Proj}_{A(I)}^{nor}$) sending a vector v outside of $\mathrm{Lsp}(A(I))$ onto the unit vector along the component of v in the orthogonal complement of $\mathrm{Lsp}(A(I))$. By agreement, vectors in $\mathrm{Lsp}(A(I))$ are sent onto the (-1)-simplex ϕ and we set $\mathrm{Proj}_{\phi}^{\mathrm{nor}} = \mathrm{Id}$. As an example $\mathrm{Proj}_{A(I)}^{\mathrm{nor}}(A(I^c))$ is a spherical (n-|I|-1)-simplex orthogonal to the face A(I) of A. The class $[\mathrm{Proj}_{A(I)}^{\mathrm{nor}}(A(I^c))]$ in $\mathcal{PS}$ will be called the interior codimensional n-|I| angle at the face A(I) of A. For this reason, we also write $\theta_A(A(I))$ in place of $\mathrm{Proj}_{A(I)}^{\mathrm{nor}}(A(I^c))$. In general, the coset of [A] in $\mathcal{PS}/\mathcal{CS}$ will be denoted by $\overline{[A]}$. When [A] is viewed as an angle (the degree is then the dimension of the angle), $\overline{[A]}$ is called the (associated) reduced angle. In these terms, the interior dihedral angle of a simplex (spherical, Euclidean or hyperbolic) at a codimensional 2 face is a spherical 1-simplex obtained through the normal projection map. Equivalently, the interior dihedral angle can be viewed as the collection of all interior unit normal vectors at a point z of a codimensional 2 face and bounded by the two incident codimensional 1 faces. Up to motions of O(k), the dihedral angle so defined is independent of the choice of z. The normal projection process is simply the extension of this viewpoint to higher dimensions.

Proposition 3.7. Let k be an (Archimedean) ordered, square root closed field. $\mathcal{PS}/\mathcal{CS}$ has the structure of a commutative, evenly graded and filtered Hopf algebra over $\mathbb{Z}$. The additive structure is induced by interior disjoint union. The multiplicative structure is induced by orthogonal join. The grading and (increasing) filtration are induced by degree (= 1 + dimension). The unit element is given by the class $\overline{[\phi]}$ of the (-1)-simplex ϕ (not to be confused with the zero element). The counit is induced by the augmentation: $\mathcal{PS}/\mathcal{PS}^+ \cong \mathbb{Z}$. The comultiplication is induced by the following formula on spherical n-simplices A in $S(k^\infty)$:

$$\overline{S\Psi}\,\overline{[A]} = \Sigma_I \overline{[A(I)]} \otimes \overline{[\theta_A(A(I))]} = \Sigma_I \overline{[A(I)]} \otimes \overline{[\mathrm{Proj}_{A(I)}^{\mathrm{nor}}(A(I^c))]} \quad (3.8)$$

where I ranges over all 2^{n+1} subsets of $\{0,\ldots,n\}$. Finally, the antipode is given by the duality involution.

Proof. The ring structure has already been checked. We next show that (3.8) is well defined. We will actually show that we have a well defined additive homomorphism:

$$S\Psi\colon \mathcal{P}S \longrightarrow \mathcal{P}S \otimes \mathcal{P}S/\mathcal{C}S \tag{3.9}$$

On spherical n-simplices A in $S(k^{\infty})$, $S\Psi$ is defined by the formula:

$$S\Psi[A] = \Sigma_I [A(I)] \otimes \overline{[\theta_A(A(I))]} = \Sigma_I [A(I)] \otimes \overline{[\mathrm{Proj}^{\mathrm{nor}}_{A(I)}(A(I^c))]} \tag{3.10}$$

where I ranges over all 2^{n+1} subsets of $\{0, \ldots, n\}$.

It is evident that the right side of (3.10) depends only on the O(k)-congruence class of A. The essential point is to show that the right side of (3.10) is additive with respect to simple subdivisions of A. The argument will be carried out in terms of $\mathrm{Proj}^{\mathrm{nor}}_{A(I)}$. The equivalent formulation in terms of interior angles will be evident to the geometrically minded. For our purpose, the right hand side of (3.10) will be rewritten as a double sum by first collecting all the terms with $|I|$ equal to some fixed i and then let $0 \leq i \leq n$. For i = 0 and n, we get $[\phi] \otimes \overline{[A]}$ and $[A] \otimes \overline{[\phi]}$ respectively. The additivity is clear for each of these terms. Moreover, these exhibit the counitary property of the augmentation map. Let $0 < i < n$ and consider the term in (3.10) corresponding to a particular subset I of $\{0, \ldots, n\}$ with $|I| = i$. Let $A = B \coprod C$ be a simple subdivision of A. We will examine the terms in the expressions for $S\Psi[B]$ and $S\Psi[C]$. We relabel the indices and assume that the simple subdivision is carried out at the point $\alpha x_0 + \beta x_n$ of $S(k^{\infty})$, $\alpha, \beta \in k^+$. There are several cases.

Case 1. $0, n \in I$. $A(I^c) = B(I^c) = C(I^c)$, $\mathrm{Lsp}(A(I)) = \mathrm{Lsp}(B(I)) = \mathrm{Lsp}(C(I))$ and A(I) is simply subdivided as $\alpha x_0 + \beta x_n$ into $B(I) \coprod C(I)$. Using additivity of the tensor product in the left variable, we have a precise accounting of the terms corresponding to such subsets I in $S\Psi[A]$, $S\Psi[B]$, and $S\Psi[C]$.

Case 2. $0, n \notin I$. $A(I) = B(I) = C(I)$ and $A(I^c)$ is simply subdivided at $\alpha x_0 + \beta x_n$ into $B(I^c) \coprod C(I^c)$. It follows that $\mathrm{Proj}^{\mathrm{nor}}_{A(I)}(A(I^c))$ is simply subdivided at $\mathrm{Proj}^{\mathrm{nor}}_{A(I)}(\alpha x_0 + \beta x_n)$ into $\mathrm{Proj}^{\mathrm{nor}}_{B(I)}(B(I^c)) \coprod \mathrm{Proj}^{\mathrm{nor}}_{C(I)}(C(I^c))$. Using additivity of the tensor product in the right variable, we have a precise accounting of the terms corresponding to such subsets I in $S\Psi[A]$, $S\Psi[B]$ and $S\Psi[C]$.

Case 3. $0 \in I$ and $n \notin I$, or $0 \notin I$ and $n \in I$. We can pair off the two possibilities. Namely, let $J \subset \{1, \ldots, n-1\}$. We then define $I = \{0\} \cup J$ and $I' = \{n\} \cup J$. Let $B = \mathrm{sccl}\{x_0, \ldots, x_{n-1}, z\}$, $C = \mathrm{sccl}\{z, x_1, \ldots, x_n\}$ where $z = \alpha x_0 + \beta x_n$. Clearly $A(I) = B(I)$ and $x_0 \in \mathrm{Lsp}(A(I))$. It follows that $\mathrm{Proj}^{\mathrm{nor}}_{A(I)}(z) = \mathrm{Proj}^{\mathrm{nor}}_{A(I)}(x_n)$ so that $\mathrm{Proj}^{\mathrm{nor}}_{B(I)}(B(I^c)) = \mathrm{Proj}^{\mathrm{nor}}_{A(I)}(A(I^c))$. Similar argument shows that $A(I') = C(I')$ and that $\mathrm{Proj}^{\mathrm{nor}}_{C(I')}(C(I'^c)) = \mathrm{Proj}^{\mathrm{nor}}_{A(I')}(A(I'^c))$. These already account for the terms in (3.10) corresponding to the subsets I and I' in the expression for $S\Psi[A]$. We are left with the sum:

$$[B(I')] \otimes \overline{[\mathrm{Proj}^{\mathrm{nor}}_{B(I')}(B(I'^c))]} + [C(I)] \otimes \overline{[\mathrm{Proj}^{\mathrm{nor}}_{C(I)}(C(I^c))]} \qquad (3.11)$$

Evidently, $B(I') = C(I)$ and $z \in \mathrm{Lsp}(B(I')) = \mathrm{Lsp}(C(I))$. We also have:

$$B(I'^c) = \mathrm{sccl}\{\ldots, x_j, \ldots, x_n\}, \ j \in \{1, \ldots, n-1\} - J, \text{ and}$$
$$C(I^c) = \mathrm{sccl}\{x_0, \ldots, x_j, \ldots\}, \ j \in \{1, \ldots, n-1\}.$$

Since $z = \alpha x_0 + \beta x_n$, $\alpha, \beta \in k^+$, we have $\mathrm{Proj}^{\mathrm{nor}}_{B(I')}(x_n) = -\mathrm{Proj}^{\mathrm{nor}}_{C(I)}(x_0)$. It follows that:

$$\mathrm{Proj}^{\mathrm{nor}}_{B(I')}(B(I'^c)) \coprod \mathrm{Proj}^{\mathrm{nor}}_{C(I)}(C(I^c)) \text{ is a lune.}$$

Since the right hand factors in (3.11) are reduced mod lunes, the sum in (3.11) is zero. We therefore have a precise accounting of the terms corresponding to the paired subsets I and I' in $S\Psi[A]$, $S\Psi[B]$ and $S\Psi[C]$.

These three cases imply $S\Psi$ extends to a well defined additive homomorphism on $\mathcal{P}S$ as indicated in (3.9). In order to reduce mod $\mathcal{C}S$ so that we can obtain $\overline{S\Psi}$ from $S\Psi$ we need to know the behavior of $S\Psi$ on $[A]$ when $A = \{\text{point}\} * A'$. For this purpose, we consider the more general case where $A = B * C$. If we view $\mathcal{P}S \otimes \mathcal{P}S/\mathcal{C}S$ as a commutative ring obtained from the tensor product of the commutative rings $\mathcal{P}S$ and $\mathcal{P}S/\mathcal{C}S$, then we assert that $S\Psi$ is a ring homomorphism; equivalently, the following diagram of maps is commutative:

$$\begin{array}{ccc} \mathcal{P}S \otimes \mathcal{P}S & \xrightarrow{S\Psi \otimes S\Psi} & (\mathcal{P}S \otimes \mathcal{P}S/\mathcal{C}S) \otimes (\mathcal{P}S \otimes \mathcal{P}S/\mathcal{C}S) \\ \downarrow * & & \downarrow (* \otimes *) \circ T \\ \mathcal{P}S & \xrightarrow{S\Psi} & \mathcal{P}S \otimes \mathcal{P}S/\mathcal{C}S \end{array} \qquad (3.12)$$

In (3.12), T denotes the map twisting the middle two factors. All the maps in (3.12) are additive. The verification of the commutativity of (3.12) may therefore be restricted to a spherical n-simplex $A = A(I)*A(I^c)$ where I is a suitable subset of $\{0,\ldots,n\}$. This simply means $< Lsp(A(I)), Lsp(A(I^c))> = 0$. If we let $J \subset I$, $K \subset I^c$, then the desired commutativity follows from the preceding orthogonality together with the following equalities:

$$\begin{aligned} \mathrm{Proj}^{nor}_{A(J\cup K)}(A((J\cup K)^c) &= \mathrm{Proj}^{nor}_{A(J\cup K)}(A(I-J)*A(I^c-K)) \\ &= \mathrm{Proj}^{nor}_{A(J)}(A(I-J)) * \mathrm{Proj}^{nor}_{A(K)}(A(I^c-K)). \end{aligned}$$

We note that $S\Psi[\text{point}] = [\text{point}]\otimes\overline{[\phi]} + [\phi]\otimes\overline{[\text{point}]} = [\text{point}]\otimes\overline{[\phi]}$. As a consequence, $S\Psi(\mathcal{C}S)$ is contained in the image of $\mathcal{C}S \otimes \mathcal{P}S/\mathcal{C}S$ in $\mathcal{P}S \otimes \mathcal{P}S/\mathcal{C}S$. This means that $S\Psi$ can be reduced mod $\mathcal{C}S$ to yield $\overline{S\Psi}$. We note in passing that this is precisely the place where we have used the assumption that $\mathcal{C}S$ is the ideal generated by [point].

We have already checked the counitary law. For the coassociative law, we again prove something stronger. Namely, we assert that the following diagram of maps is commutative:

$$\begin{array}{ccc} \mathcal{P}S & \xrightarrow{S\Psi} & \mathcal{P}S \otimes \mathcal{P}S/\mathcal{C}S \\ \downarrow{\scriptstyle S\Psi} & & \downarrow{\scriptstyle S\Psi\otimes\overline{Id}} \\ \mathcal{P}S \otimes \mathcal{P}S/\mathcal{C}S & \xrightarrow{Id\otimes\overline{S\Psi}} & \mathcal{P}S \otimes \mathcal{P}S/\mathcal{C}S \otimes \mathcal{P}S/\mathcal{C}S \end{array} \qquad (3.13)$$

In (3,13), all the maps are additive. The verification of the commutativity of (3.13) may be restricted to a spherical n-simplex A. We write the set $\{0,\ldots,n\}$ as a disjoint union of three subsets I, J and K (empty subset is allowed). After the diagonalization $S\Psi$, we have terms of the form:

$$[A(I)]\otimes\overline{[\mathrm{Proj}^{nor}_{A(I)}(A(J\cup K))]} \text{ and } [A(I\cup J)]\otimes\overline{[\mathrm{Proj}^{nor}_{A(I\cup J)}(A(K))]} \qquad (3.14)$$

We note that these are distinct terms if and only if $J \neq \phi$. Applying the map $Id\otimes\overline{S\Psi}$ to the first expression of (3.14) and applying the map $S\Psi\otimes\overline{Id}$ to the second expression of (3.14) will lead to a common expression:

$$[A(I)] \otimes \overline{[\mathrm{Proj}^{nor}_{A(I)}(A(J))]} \otimes \overline{[\mathrm{Proj}^{nor}_{A(I\cup J)}(A(K))]}.$$

We note that a little bit of geometry is involved in the attainment of the preceding expression when we apply $Id \otimes \overline{S\Psi}$; it amounts to the uniqueness of the Gram-Schmidt procedure. The commutativity is now analogous to the coassociativity in the shuffle algebra based on sets, see Sweedler [66] for more details. It is clear that the reduction mod $\mathcal{C}S$ argument of the preceding paragraph can be repeated to yield the coassociativity for the comultiplication $\overline{S\Psi}$.

We are left with the task of showing that the duality involution is the antipode. The behavior is evidently correct on $\overline{[\phi]}$. We now consider the behavior on $\overline{[A]}$ for $A = \mathrm{sccl}\{x_0, \ldots, x_n\}$, $n \geq 0$. We must show:

$$(* \circ (Id \otimes \S) \circ \overline{S\Psi})\overline{[A]} = 0 = (* \circ (\S \otimes Id) \circ \overline{S\Psi})\overline{[A]} \tag{3.15}$$

The left hand side of (3.15) is:

$$\overline{\Sigma_I [A(I) * \mathrm{Proj}^{nor}_{A(I)}(A(I^c))^{\S}]}, \; I \subset \{0, \ldots, n\}$$

In view of Proposition 3.4, this expression is 0 if we can show:

$$\mathrm{Proj}^{nor}_{A(I)}(A(I^c))^{\S} = A^{\S}(I^c) \tag{3.16}$$

(3.16) is clear if I or I^c is ϕ. In general, we may relabel the indices and assume $I = \{0, \ldots, i\}$. Let $A = \mathrm{sccl}\{x_0, \ldots, x_n\}$ with $A^{\S} = \mathrm{sccl}\{y_0, \ldots, y_n\}$. We can write $x_j = u_j + \gamma_j z_j$ with $u_j \in \mathrm{Lsp}(x_0, \ldots, x_i)$, $\gamma_j \in k^+$, $z_j \in S(k^{\infty})$ so that $< z_j, x_s > = 0$, $0 \leq s \leq i < j \leq n$. By definition, $\mathrm{Proj}^{nor}_{A(I)}(x_j) = z_j$. Clearly, $\mathrm{Lsp}(y_{i+1}, \ldots, y_n) = \mathrm{Lsp}(z_{i+1}, \ldots, z_n)$ is the orthogonal complement of $\mathrm{Lsp}(A(I)) = \mathrm{Lsp}(x_0, \ldots, x_i)$ in $\mathrm{Lsp}(A)$. From (3.1) and the definition of z_j, $j > i$, we obtain:

$$< y_s, z_t > = 0 \text{ for } s \neq t, \; < y_s, z_s > = < y_s, x_s > \gamma_s^{-1} < 0, \; i < s, t \leq n.$$

(3.16) now follows from the characterization (3.1) of the duality map. The right hand side of (3.15) is obtained from the left hand side by applying the duality involution. Q.E.D.

The proof of Proposition 3.7 showed that $\mathcal{P}S$ is actually a comodule for the Hopf algebra $\mathcal{P}S/\mathcal{C}S$ with $S\Psi$ as the structure map. We summarize the additional results:

Proposition 3.17. Let k be an (Archimedean) ordered, square root closed field. The map $S\Psi: \mathcal{P}S \longrightarrow \mathcal{P}S \otimes \mathcal{P}S/\mathcal{C}S$ as defined in (3.9), (3.10) is a ring homomorphism preserving the degree grading. $S\Psi$ is also the structure map turning $\mathcal{P}S$ into a (right) comodule for the commutative, filtered, graded (by degree) Hopf algebra $\mathcal{P}S/\mathcal{C}S$ over $\mathbb{Z}$. Each $\mathcal{P}S(i) = \Sigma_{0 \leq j \leq i} \mathcal{P}S^j$, $i \geq 0$, in the degree filtration is a subcomodule.

The map $S\Psi$ will be called the total spherical Dehn invariant. For each $i \geq 0$, the codimensional i spherical Dehn invariant $S\Psi^{(i)}$ is defined to be the partial sum in (3.10) over those subsets I of $\{0, \ldots, n\}$ with $|I| = n-i$. Since $\mathcal{P}S/\mathcal{C}S$ is evenly graded, we have:

$$S\Psi^{(2i+1)} = 0, \; i \geq 0 \quad \text{and} \quad S\Psi = \Sigma_{i \geq 0} S\Psi^{(2i)} \tag{3.18}$$

We note that $S\Psi^{(2i)}$ vanishes on polyhedra of dimension n in $S(k^\infty)$ as soon as $2i > n+1$; thus the infinite sum in (3.18) is effectively a finite sum when evaluated on elements of $\mathcal{P}S$. The following result holds:

Proposition 3.19. Let k be an (Archimedean) ordered, square root closed field. Let P and Q denote spherical polyhedra of dimension n in $S(k^\infty)$. Let i be the nonnegative integer so that $n = 2i$ or $2i-1$. Then:

(a) $S\Psi^{(2j)}([P] - [Q]) = 0$ for $j > i$.

(b) $S\Psi^{(0)}([P] - [Q]) = ([P] - [Q]) \otimes \overline{[\phi]}$ so that $S\Psi^{(0)}$ is an identification isomorphism between $\mathcal{P}S$ and $\mathcal{P}S \otimes (\mathcal{P}S/\mathcal{C}S)^0$.

(c) $S\Psi^{(2i)}([P] - [Q]) = 0$ if and only if $[P] - [Q] \in \mathcal{C}_2 S$.

Proof. (a) and (b) are evident. For (c), we have two cases.

Case 1. $n = 2i-1$. $S\Psi^{(2i)}([P] - [Q]) = [\phi] \otimes (\overline{[P] - [Q]})$ so that $S\Psi^{(2i)}$ can be identified with the projection map from $\mathcal{P}S^{2i-1}$ to $(\mathcal{P}S/\mathcal{C}S)^{2i-1}$. The kernel is $[\text{point}] * \mathcal{P}S^{2i-2} = [\text{point}] * [\text{point}] * \mathcal{P}S^{2i-3} \subset \mathcal{C}_2 S$.

Case 2. $n = 2i$. We automatically have $[P] - [Q] = [\text{point}]*[P'] - [\text{point}]*[Q']$. Since $S\Psi$ is multiplicative, it follows that:

$$S\Psi^{(2i)}([P] - [Q]) = [\text{point}] \otimes (\overline{[P'] - [Q']})$$

Since $\mathcal{P}S^1$ is infinite cyclic with [point] as a generator, the preceding expression is 0 if and only if $[P'] - [Q'] \in \mathcal{C}S$. From this we easily conclude that the preceding expression is 0 if and only if $[P] - [Q] \in \mathcal{C}_2S$. We note that in fact $\mathcal{C}_2S$ can be replaced by $\mathcal{C}_3S$ in this case. Q.E.D.

In general, the structure map for a comodule over a Hopf algebra is always injective (compare Proposition 3.19, (b)). In this respect, $S\Psi$ is not to the point as far as solving spherical scissors congruence goes. A little more to the point is to consider the following map:

$$S\Phi = (\text{gr. vol.} \otimes \overline{\text{Id}}) \circ S\Psi : \mathcal{P}S \longrightarrow \mathbb{R}[T] \otimes \mathcal{P}S/\mathcal{C}S \qquad (3.20)$$

In order to make sense out of $S\Phi$, we assume k to be Archimedean. In accordance with (3.18), we have:

$$S\Phi^{(2i+1)} = 0, \ i \geq 0 \text{ and } S\Phi = \Sigma_{i \geq 0} S\Phi^{(2i)} \qquad (3.21)$$

It is immediate that $S\Phi^{(0)}$ is just the graded volume map in disguise. $S\Phi^{(2)}$ is just the invariant introduced by Dehn. For these reasons, we will call $S\Phi$ the <u>classical</u> <u>total</u> <u>spherical</u> <u>Dehn</u> <u>invariant</u>.

<u>Proposition 3.22</u>. Let k be an Archimedean ordered, square root closed field. The classical total spherical Dehn invariant $S\Phi$ is a ring homomorphism preserving the degree grading. The graded ideal ker $S\Phi$ is equal to the graded ideal $\text{Tor}(\mathcal{P}S)$ formed by all the (additive) torsion elements of $\mathcal{P}S$. If $0 \leq i \leq 3$, then $\text{Tor}(\mathcal{P}S)^i = 0$.

<u>Proof</u>. Graded volume, $\overline{\text{Id}}$ and $S\Psi$ are all ring homomorphisms preserving the degree grading, the same therefore holds for $S\Phi$. $\mathbb{R}[T] \otimes \mathcal{P}S/\mathcal{C}S$ is an $\mathbb{R}$-vector space through the left factor, it follows that $\text{Tor}(\mathcal{P}S) \subset \ker S\Phi$. For $0 \leq i \leq 3$, $S\Phi^{(0)}$ is already injective on $\mathcal{P}S^i$. We therefore have:

$$\text{Tor}(\mathcal{P}S)^i = 0 = (\ker S\Phi)^i, \ 0 \leq i \leq 3.$$

Since $\text{Tor}(\mathcal{P}S)$ and ker $S\Phi$ are both graded ideals, we only need to show $(\ker S\Phi)^{n+1} \subset \text{Tor}(\mathcal{P}S)$ for each $n \geq 3$. Let $[P] - [Q] \in (\ker S\Phi)^{n+1}$. As in Proposition 3.19, let $i = 1 + [n/2]$. From the proof of (c), Proposition 3.19, we have:

$$0 = S\Phi^{(2i)}([P] - [Q]) = 1 \otimes ([P] - [Q]) \in \mathbb{R} \otimes (\mathcal{P}S/\mathcal{C}S) \qquad \text{if } n = 2i-1$$
$$= T \otimes ([P'] - [Q']) \in \mathbb{R}T \otimes (\mathcal{P}S/\mathcal{C}S) \qquad \text{if } n = 2i.$$

Applying the functor $\otimes(\mathcal{P}S/\mathcal{C}S)$ to the exact sequence:

$$0 \longrightarrow \mathbb{Z} \longrightarrow \mathbb{R} \longrightarrow \mathbb{R}/\mathbb{Z} \longrightarrow 0$$

where $1 \in \mathbb{R}$ corresponds to an appropriate power of T, it is immediate that $[P] - [Q]$ respectively $[P'] - [Q']$ lies in $\mathrm{Tor}(\mathcal{P}S/\mathcal{C}S)$. As in Proposition 3.19, we can find $N_1 \in \mathbb{Z}^+$ and $[P''] - [Q''] \in \mathcal{P}S^{n-1}$ so that:

$$N_1([P] - [Q]) = [\text{point}] * [\text{point}] * ([P''] - [Q'']) \tag{3.23}$$

Applying $S\Phi$ to both sides and using the multiplicativity of $S\Phi$, we have:

$$0 = (T^2 \otimes \overline{[\phi]}) * S\Phi^{(2i-2)}([P''] - [Q''])$$

Similar reasoning shows that $[P''] - [Q''] \in \mathrm{Tor}(\mathcal{P}S/\mathcal{C}S)^{n-1}$. Evidently, our argument can be repeated. In a finite number of steps, we can find $N \in \mathbb{Z}^+$ so that:

$$N([P] - [Q]) \in \mathcal{C}_{2i}S \cap \mathcal{P}S^{n+1} \subset \mathcal{P}S_n^{n+1}.$$

Since $0 = S\Phi^{(0)}([P] - [Q])$ and the graded volume map $S\Phi^{(0)}$ is injective on $\mathcal{P}S_n^{n+1}$ (cf. Proposition 2.6), we have: $N([P] - [Q]) = 0$. Q.E.D.

In order to have a better picture of the map $S\Phi$, we consider the case $k = \mathbb{R}$. On $\mathcal{P}S^0$ and $\mathcal{P}S^1$, $S\Phi$ is just $S\Phi^{(0)}$ and leads to $\mathcal{P}S^0 \cong \mathcal{P}S^1 \cong \mathbb{Z}$. On $\mathcal{P}S^2$ and $\mathcal{P}S^3$, $S\Phi^{(0)}$ leads to $\mathcal{P}S^2 \cong \mathcal{P}S^3 \cong \mathbb{R}$ while $S\Phi^{(2)}$ amounts to reduction mod $\mathbb{Z}$ of $S\Phi^{(0)}$ so that $S\Phi$ does not carry more information than graded volume. On $\mathcal{P}S^4$, $S\Phi$ begins to yield more information than $S\Phi^{(0)}$. We recall the join filtration:

$$\mathcal{P}S^4 = \mathcal{P}S_1^4 \supset \mathcal{P}S_2^4 \supset \mathcal{P}S_3^4 \supset \mathcal{P}S_4^4 \supset \mathcal{P}S_5^4 = \ldots = 0.$$

The graded volume $S\Phi^{(0)}$ leads to $\mathcal{P}S_3^4 \cong \mathbb{R}$ and carries $\mathcal{P}S_4^4$ onto $\mathbb{Z}$. This leads to $\mathcal{P}S^4/\mathcal{P}S_3^4 \cong \ker S\Phi^{(0)} \cap \mathcal{P}S^4$. In order to simplify the notation, we identify $\mathcal{P}S_3^4$ with $\mathbb{R}$ through $S\Phi^{(0)}$. Similarly, we identify $\mathcal{P}S^2$ with $\mathbb{R}$ by means of $S\Phi^{(0)}$. Specifically, $[\mathrm{sccl}\{x_0, x_1\}]$ is identified with the real number α with $0 < \alpha < 2$ so that $< x_0, x_1 > = \cos(\pi\alpha/2)$. We let $\overline{\alpha}$ denote the coset of α in $\mathbb{R}/\mathbb{Z}$. With this notation, $\wp S_3^4$ can be identified with $1*\alpha$, $\alpha \in \mathbb{R}$ and $\wp S_2^4$ is generated by all $\alpha * \beta$ with $0 < \alpha, \beta < 2$. The symbol $\alpha*\beta$ is biadditive and symmetric in α, β.

Proposition 3.24. Let k be an Archimedean ordered, square root closed field. Then $\ker S\Phi^{(0)} \cap \ker S\Phi^{(2)} \cap \wp S_2^4 = 0$ and $\wp S_2^4$ is isomorphic to the degree 2 component $S_{\mathbb{Z}}^2(\wp S^2)$ of the symmetric algebra over $\mathbb{Z}$ based on the torsion free abelian group $\wp S^2$.

Proof. We first consider the case $k = \mathbb{R}$. It is clear that the symbol $\alpha*\beta$ leads to a surjective map from $S_{\mathbb{Z}}^2(\mathbb{R})$ to $\wp S_2^4$ where $\wp S^2$ has been identified with $\mathbb{R}$. Since $\mathbb{R}$ is a $\mathbb{Q}$-vector space, $S_{\mathbb{Z}}^2(\mathbb{R})$ is the same as $S_{\mathbb{Q}}^2(\mathbb{R})$. We can write $\mathbb{R} = \mathbb{Q} \coprod W$ for a $\mathbb{Q}$-subspace W of $\mathbb{R}$. We then have:

$$S_{\mathbb{Q}}^2(\mathbb{R}) \cong S_{\mathbb{Q}}^2(W) \coprod W \coprod \mathbb{Q} \tag{3.25}$$

Under the homomorphism to $\wp S_2^4$, the summand $W \coprod \mathbb{Q} = \mathbb{R}$ is mapped onto $\wp S_3^4$ with $\alpha \in \mathbb{R}$ sent onto $1*\alpha = \alpha*1$. In general, if V is any vector space over a field F of characteristic not 2, $S_F^2(V)$ can be identified with the subspace $(V \otimes_F V)^+$ of the symmetric tensors (invariant elements under the twisting map T). The element $v \cdot w$ in $S_F^2(V)$, $v, w \in V$, is identified with the element $(v \otimes w + w \otimes v)/2$ in $(V \otimes_F V)^+$.

With our notations, $S\Phi^{(2)}(\alpha*\beta) = \alpha T^2 \otimes \bar{\beta} + \beta T^2 \otimes \bar{\alpha}$. If we reduce both α and β mod $\mathbb{Z}$ and use the isomorphisms: $S_{\mathbb{Z}}^2(\mathbb{R}/\mathbb{Z}) \cong S_{\mathbb{Q}}^2(\mathbb{R}/\mathbb{Q}) \cong S_{\mathbb{Q}}^2(W)$, then $S\Phi^{(2)}$ induces an isomorphism between $\wp S_2^4/\wp S_3^4$ and $S_{\mathbb{Q}}^2(W)$ corresponding to the projection map in (3.25). This shows that $\ker S\Phi^{(2)} \cap \wp S_2^4 \subset \wp S_3^4$. Since $S\Phi^{(0)}$ in injective on $\wp S_3^4$, we have $\ker S\Phi^{(0)} \cap \ker S\Phi^{(2)} \cap \wp S_2^4 = 0$. Furthermore, $S\Phi^{(0)}$ defines $\wp S_3^4 \cong \mathbb{R}$ corresponding to the identification of $\mathbb{R}$ with the component $W \coprod \mathbb{Q}$ in (3.25). From this follows without much difficulty the isomorphism between $\wp S_2^4$ and $S_{\mathbb{Z}}^2(\mathbb{R})$ induced by sending α, β in $\mathbb{R}$ with $0 < \alpha, \beta < 2$ onto $\alpha*\beta$.

The general case follows from the case $k = \mathbb{R}$. We recall that $\wp S^2$ is torsion free and 2-divisible, and $S\Phi^{(0)}$ identifies $\wp S^2$ with an additive subgroup of $\mathbb{R}$. We omit further details. Q.E.D.

In view of (3.20), $S\Phi$ carries less information than $S\Psi$. Since $S\Psi^{(0)}$ is an identification isomorphism, we can ask: How much information is carried by $S\Psi^{(2i)}$, $i > 0$? In general, if $\omega: M \longrightarrow M \otimes C$ is the structure map of a right comodule M for the Hopf algebra C, then the C-invariant elements on M is defined by the formula:

$$M^C = \{x \in M \mid \omega(x) = x \otimes 1\}$$

Through the use of derived functors, this leads to cohomology groups $H^i(C, M)$, $i \geq 0$, where $H^0(C, M) = M^C$. With $S\Psi$ as the structure map, we immediately have:

$$H^0(\mathcal{P}S/\mathcal{C}S, \mathcal{P}S) = \bigcap_{i>0} \ker S\Psi^{(2i)}$$

Since $S\Psi$ is also a ring homomorphism, the subring $\mathbb{Z}[[\text{point}]]$ of $\mathcal{P}S$ generated by $[\text{point}]$ is contained in $H^0(\mathcal{P}S/\mathcal{C}S, \mathcal{P}S)$.

Proposition 3.26. Let k be an ordered, square root closed field.

(a) $H^0(\mathcal{P}S/\mathcal{C}S, \mathcal{P}S) \subset \mathrm{Tor}(\mathcal{C}_2 S \bmod \mathbb{Z}[[\text{point}]]) + \mathbb{Z}[[\text{point}]]$;

(b) If k is Archimedean, then $H^0(\mathcal{P}S/\mathcal{C}S, \mathcal{P}S) \subset \mathrm{Tor}(\mathcal{C}_2 S) + \mathbb{Z}[[\text{point}]]$, $\mathbb{Z}[[\text{point}]] \cong \mathbb{Z}[T]$; and

(c) If k is Archimedean and $\mathcal{P}S^2$ is divisible (equivalently, $SO(2;k)$ is divisible), then $H^0(\mathcal{P}S/\mathcal{C}S, \mathcal{P}S) = \mathbb{Z}[[\text{point}]] \cong \mathbb{Z}[T]$.

Proof. $H^0(\mathcal{P}S/\mathcal{C}S, \mathcal{P}S)$ is a graded subgroup of $\mathcal{P}S$ because each $S\Psi^{(2i)}$ is a graded homomorphism. Let P, Q be spherical polyhedra of dimension n in $S(k^\infty)$ so that $[P] - [Q] \in H^0(\mathcal{P}S/\mathcal{C}S, \mathcal{P}S)$. Adding on suitable spherical polyhedron R to P and Q, we may assume $[Q] \in \mathbb{Z}[[\text{point}]]$. Since we already know $\mathbb{Z}[[\text{point}]] \subset H^0(\mathcal{P}S/\mathcal{C}S, \mathcal{P}S)$, we may assume $[P]$ is in $H^0(\mathcal{P}S/\mathcal{C}S, \mathcal{P}S)$. As in Proposition 3.19, we can assume $[P]$ to have form: $[\text{point}] * [\text{point}] * [P'']$. If $i = 1 + [n/2]$, then we have:

$$S\Psi^{(2i-2)}([P]) = ([\text{point}] * [\text{point}]) \otimes \overline{[P'']}$$

According to the torsion structure of $\mathcal{P}S^2/\mathcal{P}S^2_2 \cong SO(2;k)$, we can find an integer $N_2 > 0$ so that $N_2[P''] \in \mathcal{C}S$. The argument now proceeds as in the proof of Proposition 3.22. In a finite number of steps, we can find an integer $N > 0$ so that $N[P''] \in \mathbb{Z}[[\text{point}]]$. (a) therefore holds. When k is Archimedean, $\ker S\Phi^{(0)} \cap H^0(\mathcal{P}S/\mathcal{C}S, \mathcal{P}S) \subset \bigcap_{i \geq 0} \ker S\Phi^{(2i)} = \ker S\Phi$ because $S\Phi^{(2i)} = (\text{gr. vol.} \otimes \overline{\text{Id}}) \circ S\Psi^{(2i)}$ for $i \geq 0$. Since $S\Phi^{(0)}$ induces the isomorphism between $\mathbb{Z}[[\text{point}]]$ and $\mathbb{Z}[T]$, (b) follows from (a). Suppose that k is Archimedean and $\mathcal{P}S^2$ is divisible. In view of (b), it is enough to show that $\text{Tor}(\mathcal{C}_2 S) \cap H^0(\mathcal{P}S/\mathcal{C}S, \mathcal{P}S) = 0$. If we examine the proof of (a), we see that an element of $\text{Tor}(\mathcal{C}_2 S)^n \cap H^0(\mathcal{P}S/\mathcal{C}S, \mathcal{P}S)$ has the form:

$$[\text{point}] * [\text{point}] * ([P''] - [Q'']) \tag{3.27}$$

where $N([P''] - [Q'']) \in \mathbb{Z}[[\text{point}]]$. Since $\text{Tor}(\mathcal{P}S) \subset \ker S\Phi^{(0)}$ and $S\Phi^{(0)}$ is injective on $\mathbb{Z}[[\text{point}]]$, $N([P''] - [Q'']) = 0$. The divisibility of $\mathcal{P}S^2$ forces the expression in (3.27) to be 0. Q.E.D.

Remark 3.28. The proof of (c) in Proposition 3.26 only used the divisibility of $[\text{point}] * [\text{point}]$. When k is Archimedean, the invariant $S\Phi$ separates $\mathcal{P}S$ modulo possible torsions in $\mathcal{P}S$ while the invariants $S\Phi^{(0)}$ and $S\Psi^{(2i)}$, $i > 0$, may separate $\mathcal{P}S$ modulo a smaller subgroup of $\text{Tor}(\mathcal{P}S)$. It is not known if $\text{Tor}(\mathcal{P}S)$ could be nontrivial.

Theorem 3.29. Let k be an Archimedean ordered, square root closed field. $\mathcal{P}S/\text{Tor}(\mathcal{P}S)$ and $(\mathcal{P}S/\mathcal{C}S)/\text{Tor}(\mathcal{P}S/\mathcal{C}S)$ are both integral domains. In particular, if P, Q, R are nonempty spherical polyhedra in $S(k^\infty)$ so that $< \text{Lsp}(P) + \text{Lsp}(Q), \text{Lsp}(R) > = 0$, then $P * R$ is scissors congruent to $Q*R$ if and only if there is a positive integer N such that:

> $\coprod_i P_i$ is scissors congruent to $\coprod_i Q_i$, $1 \leq i \leq N$, where
> P_i, Q_i are respectively congruent to P, Q for each i.

The proof is based on the Hopf-Leray theorem on the algebra structure of a commutative, graded, connected, Hopf algebra over a field of characteristic 0. More details can be found in Appendix: Propositions 1.6 and 1.7. The geometric assertion is an easy corollary of the algebraic assertion that $\mathcal{P}S/\text{Tor}(\mathcal{P}S)$ is an integral domain. We do not know of any direct proof of this geometric assertion.

We will now rephrase the results contained in various propositions:

Theorem 3.30. Let k be an Archimedean ordered, square root closed field. The classical total spherical Dehn invariant $S\Phi: \mathcal{P}S \longrightarrow \mathbb{R}[T] \otimes \mathcal{P}S/\mathcal{C}S$ has kernel $\mathrm{Tor}(\mathcal{P}S)$. The image of $S\Phi$ is a graded subring such that it generates $\mathbb{R}[T] \otimes \mathcal{P}S/\mathcal{C}S$ as an $\mathbb{R}$-algebra. If $A(2i;\lambda)$, $\lambda \in \Lambda_{2i}$, are spherical (2i-1)-simplices so that $1 \otimes \overline{[A(2i;\lambda)]}$ form a $\mathbb{Q}$-basis for $\mathbb{Q} \otimes \mathcal{P}S^{2i}/\mathcal{P}S_2^{2i}$, then $\mathbb{R}[T] \otimes \mathcal{P}S/\mathcal{C}S$ is isomorphic to the polynomial algebra over $\mathbb{R}$ with generators corresponding to the images of [point] and all $[A(2i;\lambda)]$, $i > 0$, $\lambda \in \Lambda_{2i}$.

Proof. The assertion about the kernel of $S\Phi$ had been verified in Proposition 3.22. According to the theorem of Hopf-Leray, $\mathbb{R}[T] \otimes \mathcal{P}S/\mathcal{C}S$ is a polynomial algebra over $\mathbb{R}$ based on the $\mathbb{R}$-vector space of indecomposable elements. The space of indecomposable elements is just the direct sum of $\mathbb{R}T \otimes \overline{[\phi]}$ and $\coprod_{i>0} \mathbb{R} \otimes \mathcal{P}S^{2i}/\mathcal{P}S_2^{2i}$. $S\Phi$ carries [point] onto $T \otimes \overline{[\phi]}$. Modulo decomposable elements built out of $\mathbb{R}$-linear combinations of the images of [point] and $[A(2j;\mu)]$, $0 < j < i$, $\mu \in \Lambda_{2j}$, the image of $[A(2i;\lambda)]$ is just the element $1 \otimes \overline{[A(2i;\lambda)]}$. Our assertion follows through induction. Q.E.D.

Remark 3.31. Modulo the difficult questions posed by torsion and divisibility, the preceding theorem is somewhat analogous to the nonconstructive solution of the scissors congruence problem for Euclidean spaces of all dimensions given in Hadwiger's book [31; Theorem 8, p. 58]. In case k is $\mathbb{R}$, the situation is slightly clearer. $\mathcal{P}S^2$ is isomorphic to $\mathbb{R}$ through the volume map. Since $\mathcal{P}S^2$ is always torsionfree, $\mathcal{P}S^3 \cong \mathcal{P}S^2$ through the orthogonal join multiplication by [point]. $\mathcal{P}S_2^4 \cong (\mathcal{P}S^2 \otimes \mathcal{P}S^2)^+ \cong (\mathbb{R} \otimes \mathbb{R})^+$ as in Proposition 3.24. With these isomorphisms in view, we can introduce an $\mathbb{R}$-vector space structure on $\mathcal{P}S^2$ and $\mathcal{P}S_2^4$ so that $S\Phi$ defines $\mathbb{R}$-vector space injections on these groups. We do not know how such a process can be carried out in general. In particular, we do not have the spherical analogue of the canonical decompositions, nor the analogue of Theorem 3.1.1. The common theme behind these missing ingredients is the absence of an analogue of the similarity maps. In terms of the ray model, the similarity maps induce the identity map on spherical spaces.

The classical spherical Dehn invariants $S\Phi^{(2i)}$, $i \geq 0$, differ from the affine Hadwiger invariants $\Omega^n(r_1, \ldots, r_j)$ in two fundamental aspects. First, $S\Phi^{(2i)}$, $i > 0$, are not "lined up" with the join filtration. Second, $S\Phi^{(2i)}$, $i > 0$, are not "numerical".

In a private communication, B. Jessen and A. Thorup suggested the following modification of $S\Phi^{(2)}$. When k is Archimedean, $(\mathcal{P}S/\mathcal{C}S)^2 \cong SO(2;k)$ is a subgroup of $SO(2;\mathbb{R}) \cong \mathbb{R}/\mathbb{Z}$. Classically, $S\Phi^{(0)}$ and $S\Phi^{(2)}$ are defined to have values in $\mathbb{R}$ and $\mathbb{R} \otimes \mathbb{R}/\mathbb{Z}$ respectively. By ignoring the powers of T, the map $S\Phi^{(2)} - 1 \otimes \overline{S\Phi}^{(0)}: \mathcal{P}S \longrightarrow \mathbb{R} \otimes \mathbb{R}/\mathbb{Z}$ is then an additive homomorphism. Combining the multiplicative character of $S\Phi$ with the Gauss-Bonnet formula, this map vanishes on $(\mathcal{C}_{i-2}S)^i$. When i = 4, it vanishes on $\mathcal{P}S_3^4 = (\mathcal{C}_2 S)^4 = (\mathcal{C}S)^4$. The argument used in the proof of Proposition 3.24 shows that this map induces an injection on $\mathcal{P}S_2^4/\mathcal{P}S_3^4$. It is clear that:

$$\ker S\Phi^{(0)} \cap \ker S\Phi^{(2)} = \ker S\Phi^{(0)} \cap \ker(S\Phi^{(2)} - 1 \otimes \overline{S\Phi}^{(0)})$$

In the sense of separating points of $\mathcal{P}S$, we neither gain nor lose grounds by using the pair $S\Phi^{(0)}$ and $S\Phi^{(2)} - 1 \otimes \overline{S\Phi}^{(0)}$ in place of $S\Phi^{(0)}$ and $S\Phi^{(2)}$. The main problem for i = 4 is that we do not know if $S\Phi^{(2)} - 1 \otimes \overline{S\Phi}^{(0)}$ is injective on $\mathcal{P}S^4/\mathcal{P}S_3^4 \cong (\mathcal{P}S/\mathcal{C}S)^4$. This is equivalent with the following question:

Is it true that $\mathcal{P}S^4$ is torsion free and that:

$$\mathcal{P}S^4 \cap \ker S\Phi^{(0)} \cap \ker S\Phi^{(2)} \subset \ker S\Phi^{(4)} \;?$$

Aside from problems of number theoretic nature, the invariant $S\Phi^{(2)}$ can be used to generate numerical invariants. Specifically, we have:

$$\mathrm{Hom}_{\mathbb{R}}(\mathbb{R} \otimes \mathbb{R}/\mathbb{Z}, \mathbb{R}) \cong \mathrm{Hom}_{\mathbb{Z}}(\mathbb{R}/\mathbb{Z}, \mathbb{R})$$

where $\mathbb{R} \otimes \mathbb{R}/\mathbb{Z}$ is viewed as an $\mathbb{R}$-vector space through the left factor. By composing $S\Phi^{(2)}$ with elements of $\mathrm{Hom}_{\mathbb{Z}}(\mathbb{R}/\mathbb{Z}, \mathbb{R})$, we obtain numerical invariants corresponding to elements of $\mathrm{Hom}_{\mathbb{Z}}(\mathcal{P}S, \mathbb{R})$. With this in mind, Hadwiger [31; Chapter 2] indicated a generalization of Dehn invariants for the Euclidean case. The procedure is quite formal. The target group of the invariants are of the form $\mathbb{R} \otimes (\mathbb{R}/\mathbb{Z})^{\otimes i}$. We will now extend this idea in the setting of Hopf algebras.

Let $2i$ be an even integer with $i \geq 0$. We will define Hopf algebra homomorphisms into various shuffle algebras (see Appendix for references):

$$\xi_{2i}: \mathcal{P}S/\mathcal{C}S \longrightarrow Sh_{\mathbb{Z}}(\mathcal{P}S^{2i}/\mathcal{P}S_2^{2i}) \tag{3.32}$$

ξ_0 is simply the augmentation map with image $\mathbb{Z} = Sh^0_{\mathbb{Z}}(\mathcal{P}S^0)$. For $i > 0$, the basic elements of $Sh_{\mathbb{Z}}(\mathcal{P}S^{2i}/\mathcal{P}S_2^{2i})$ are of the form: $(\alpha_1, \ldots, \alpha_t)$ where α_j is the coset mod $\mathcal{P}S_2^{2i}$ of the class $[A_j]$ for a suitable spherical $(2i-1)$-simplex A_j. We will view $[A_j]$ as an angle θ_j and its coset mod $\mathcal{P}S_2^{2i}$ will be denoted by $\overline{\theta}_j$. ξ_{2i} will be defined on the level of $\mathcal{P}S^n$. For $n = 0$, $\xi_{2i}[\phi] = (.)$ and extends to an isomorphism between $(\mathcal{P}S/\mathcal{C}S)^0$ and $Sh^0_{\mathbb{Z}}(\mathcal{P}S^{2i}/\mathcal{P}S_0^{2i})$. For $n > 0$ and n not divisible by $2i > 0$, ξ_{2i} is 0 on $\mathcal{P}S^n$. Let $n = 2it > 0$, i, t in $\mathbb{Z}^+$. Let $A^{(j)}$ denote a typical codimensional j face of a spherical $(n-1)$-simplex A. For each codimensional $2i$ face sequence of length t:

$$A = A^{(0)} \supset A^{(2i)} \supset \ldots \supset A^{(2it)} = \phi$$

we associate an angle sequence $\theta_A(A^{(2it)}, \ldots, A^{(2i)}) = (\overline{\theta}_t, \ldots, \overline{\theta}_1)$ where θ_j is the interior angle at the codimensional $2i$ face $A^{(2ij-2i)}$ of $A^{(2ij)}$. More precisely, let $A^{(2ij)} = \text{sccl}\{u_1, \ldots, u_{2ij}\}$ with $A^{(2ij-2i)} = \text{sccl}\{u_{2i+1}, \ldots, u_{2ij}\}$, then $\theta_j = [\text{sccl}\{v_1, \ldots, v_{2i}\}]$ where $v_s = \text{Proj}^{\text{nor}}_A(2ij-2i)(u_s)$ is the unit vector along the normal component of u_s to $\text{Lsp}(A^{(2ij-2i)})$. We define $\xi_{2i}[A]$ to be:

$\Sigma\, \theta_A(A^{(2it)}, \ldots, A^{(2i)})$, where the summation extends over all codimensional $2i$ face sequences of length t starting from A.

If ξ^n_{2i} denotes the restriction of ξ_{2i} to $\mathcal{P}S^n$, then $\xi^n_{2i} = (\xi^{n-2i}_{2i} \otimes \text{Id}) \circ S\Psi^{(2i)}$. This observation together with induction shows that we have an additive homomorphism in (3.32).

We next check the multiplicativity of ξ_{2i}. We may assume $i > 0$. Consider an orthogonal join $P*Q$ of spherical $(p-1)$- and $(q-1)$-simplices P and Q. A typical codimensional $2i$ face has the form $P^{(s)}*Q^{(t)}$ with $s+t = 2i$, $s, t \geq 0$. Unless s or t is 0, the interior angle θ at such a face lies in $\mathcal{P}S_2^{2i}$. When s or $t = 0$, we get the interior angle $\theta_Q(Q^{(2i)})$ and $\theta_P(P^{(2i)})$ respectively. It is now evident that $\xi_{2i}[P*Q]$ is precisely the shuffle product of $\xi_{2i}[P]$ and $\xi_{2i}[Q]$. In particular, $\xi_{2i}[P*Q]$ is 0 when p or q is not a multiple of $2i$.

It remains for us to check the comultiplicativity of ξ_{2i}. We can restrict our attention to $[A]$ where A is a spherical (2it-1)-simplex, $t \in \mathbb{Z}^+$. Then:

$$(\xi_{2i} \otimes \xi_{2i}) \circ \overline{S\Psi}\,\overline{[A]} = \Sigma_I\, \xi_{2i} \overline{[A(I)]} \otimes \overline{[\theta_A(A(I))]}$$

where the summation extends over all subsets I of $\{0, \ldots, 2it-1\}$ (including $\emptyset$) with $|I|$ divisible by 2i. When we write out the angle sequences for a general term in the preceding sum, we get the expression:

$$(\overline{\theta}_t, \ldots, \overline{\theta}_{j+1}) \otimes (\overline{\theta}_j, \ldots, \overline{\theta}_1)$$

where $(\overline{\theta}_t, \ldots, \overline{\theta}_1)$ is the angle sequence associated to a specific codimensional 2i face sequence of length t starting from A, running through A(I) and ending at $\emptyset$. In the shuffle algebra $Sh_{\mathbb{Z}}(\mathcal{P}S^{2i}/\mathcal{P}S_2^{2i})$, the comultiplication carries $(\overline{\theta}_t, \ldots, \overline{\theta}_1)$ onto the following sum:

$$(.)\otimes(\overline{\theta}_t, \ldots, \overline{\theta}_1) + \ldots + (\overline{\theta}_t, \ldots, \overline{\theta}_{j+1}) \otimes (\overline{\theta}_j, \ldots, \overline{\theta}_1) + \ldots + (\overline{\theta}_t, \ldots, \overline{\theta}_1) \otimes (.)$$

The comultiplicativity of ξ_{2i} follows from counting of terms.

We have therefore shown that ξ_{2i} is a Hopf algebra homomorphism for $i \geq 0$. In passing, the antipode of $Sh_{\mathbb{Z}}(\mathcal{P}S^{2i}/\mathcal{P}S_2^{2i})$ is the involution carrying $(\overline{\theta}_t, \ldots, \overline{\theta}_1)$ onto $(-\overline{\theta}_1, \ldots, -\overline{\theta}_t) = (-1)^t(\overline{\theta}_1, \ldots, \overline{\theta}_t)$. This exhibits the compatibility with the duality involution in $\mathcal{P}S/\mathcal{C}S$ in a transparent manner. In general, $\mathcal{P}S/\mathcal{C}S$ is not cocommutative. As a result, ξ_{2i} can not be combined in the sense of tensor product to yield a Hopf algebra homomorphism. However, it is possible to combine them to form an algebra homomorphism from $\mathcal{P}S/\mathcal{C}S$ into $\otimes_i Sh_{\mathbb{Z}}(\mathcal{P}S^{2i}/\mathcal{P}S_2^{2i})$. For this, we use the fact that $\mathcal{P}S/\mathcal{C}S$ is a Hopf algebra and the fact that ξ_{2i} is 0 on $\mathcal{P}S^n$ when $0 < n < 2i$. When $i > 1$, ξ_{2i} have large kernels. It is not known how these kernels are related to each other. As a consequence, the precise kernel of the combined algebra homomorphism is unknown. On the other hand, ξ_{2i} can be combined with $S\Psi$ or $S\Phi$. For example, when k is Archimedean, $(Id \otimes \xi_0) \circ S\Phi$ is essentially the graded volume map $S\Phi^{(0)}$ while $(Id \otimes \xi_2) \circ S\Phi$ is the <u>total</u> <u>classical</u> <u>spherical</u> <u>Dehn</u>-<u>Hadwiger</u> <u>invariant</u>. In the Appendix, we use ξ_2 to get an idea of the size of $\mathcal{P}S$ as a ring.

4. OPEN PROBLEMS.

We begin by rephrasing the scissors congruence problem.

Strong Scissors Congruence Problem. (Spherical Case.) Let k denote an (Archimedean) ordered, square root closed field.

Full ring version. Find enough reasonable ring homomorphisms (respecting all structures) to separate the points of $\mathcal{P}S$.

Even subring version. Find enough reasonable ring homomorphisms (respecting all structures) to separate the points of $\mathcal{P}S^{even} = \coprod_i \mathcal{P}S^{2i}$.

Hopf algebra version. Find enough reasonable Hopf algebra homomorphisms (respecting all structures) to separate the points of $\mathcal{P}S/\mathcal{C}S$. In particular, determine the kernel of ξ_{2i} for $i > 0$.

We note that there is not too much difference between the even subring version and the full ring version. More precisely, $\mathcal{P}S^{odd}$ is a cyclic module for $\mathcal{P}S^{even}$ with [point] as a generator. When k is Archimedean, the annihilator of [point] is part of $Tor(\mathcal{P}S^{even}) = Tor(\mathcal{P}S)^{even}$. It is natural to ask:

Combinatorial Gauss-Bonnet Problem. (Spherical Case.) Is there an additive homomorphism:

$$e : \mathcal{P}S^{2i+1} \longrightarrow \mathcal{P}S^{2i}, \ i \geq 0,$$

such that e is the inverse to $*$[point] ? Equivalently, is $*$[point] an isomorphism between $\mathcal{P}S^{2i}$ and $\mathcal{P}S^{2i+1}$, $i \geq 0$?

For comparison, see section 1 of Chapter 8. In a sense, the preceding problem is connected with the ring structure of $\mathcal{P}S$. We pose two more.

Divisibility Problem. Let k be an (Archimedean) ordered, square root closed field. When is $\mathcal{P}S^i$ divisible, $i \geq 2$? Specifically, assume $k = \mathbb{R}$, is $\mathcal{P}S^i$ divisible for $i \geq 4$?

Torsion Problem. Let k be an (Archimedean) ordered, square root closed field. When is $\mathcal{P}S^i$ torsionfree, $i \geq 4$?

In contrast to the affine case, we should not expect to have a k-vector space structure on $\mathcal{P}S^i$, $i \geq 2$, even if we succeed in showing divisibility and torsion freeness in a fairly general setting, compare Propositions 2.1 and 4.1. Both of these problems seem to have something to do with the analytic functions involved in volume calculations (when k is Archimedean).

As an illustration of possible torsion, we consider the case $k = \mathbb{R}$. Let G_j, $j = 1, 2$, be finite subgroups of the same order N in $O(n+1;\mathbb{R})$. Following the classical argument of Dirichlet in constructing fundamental domains, we can find spherical polyhedral fundamental domains P_j for G_j on the n-sphere $S(\mathbb{R}^{n+1})$. According to a simple argument of Siegel [65; Lemma 3], the scissors congruence class of P_j depends only on G_j and its conjugacy class in $O(n+1;\mathbb{R})$. It is evident that $N([P_1] - [P_2]) = 0$. However, it is not evident that $[P_1] = [P_2]$. We can assume that G_1 is cyclic and that it is contained in $O(2;\mathbb{R})$ and embedded in $O(n+1;\mathbb{R})$ in the trivial manner. We already run into open problems when $n = 3$. As far as we know, the problem already exists when P_2 is a fundamental domain for the wellknown dodecahedral space. By using the obvious symmetries, this fundamental domain can be subdivided into spherical 3-simplices which are pairwise congruent under $O(4;\mathbb{R})$. The entire 3-sphere is then made up from 14400 such 3-simplices. Each such simplex is in fact the fundamental domain of a finite Coxeter group of type H_4. It is unknown if such a 3-simplex is scissors congruent to a lune. Such types of examples are finite in number for each dimension because $O(n+1;\mathbb{R})$ has only a finite number of conjugacy classes of finite groups of a fixed order N. In fact, a theorem of Jordan asserts that a finite subgroup of $GL(m;\mathbb{C})$ must have a normal abelian subgroup of index bounded by an integer depending only on m. The case of abelian finite groups can be settled--they do not lead to torsion. However, it is not clear how this can be used to settle the general case. Moreover, it is not clear if the general torsion problem can be reduced to the present situation.

The scissors congruence problem originated in the study of volume. The next few problems concern volume.

Generalized Volume Problem. Let k be an ordered, square root closed field. Does there exist a good theory of volume for spherical polyhedra? More specifically, is there an ordered field extension K of k and a graded ring homomorphism μ from $\mathcal{P}S$ to K[T] so that it "functorially" extends the definition of volume in the Archimedean case?

In a vague sense, the preceding problem can be viewed as a problem involving logic and/or nonstandard analysis. In dimension 1, the exact sequence (1.1) may be viewed as a step in this direction. Here the point is that we have a commutative group SO(2;k) which is the k-rational points of an affine algebraic group (defined over $\mathbb{Z}$).

Archimedean Volume Problem. Let k be an Archimedean ordered, square root closed field. What is the image of the graded volume map as a function of k? More generally, study the analytic and number theoretic properties of the associated volume function of an n-simplex.

From the work of L. Schläfli [63; vol. 1, 227-302], the volume of a spherical n-simplex in $S(\mathbb{R}^{n+1})$ can be expressed as an analytic function of the $n(n+1)/2$ codimensional 2 dihedral angles. When $n \geq 3$, the associated analytic functions are still quite mysterious. For a recent treatment, see Aomoto [4]. When n = 3, the dilogarithm function appears, see also the papers of Bloch [8], Coxeter [22] and the book by Lewin [49]. Special cases of the volume calculations had been made by Schläfli. These were mostly connected with the fundamental domains of finite Coxeter groups. A short summary of the work of Schläfli can be found in Coxeter [24; 142-144]. In view of these calculations, the following problem is quite natural:

Rational Simplex Problem. Let $S(\mathbb{R}^{n+1})$ have normalized volume 2^{n+1}. Let $n \geq 3$. Does there exist a spherical n-simplex with all codimensional 2 dihedral angles rational but with volume irrational?

For n < 3, the analogous problem is negatively settled by Proposition 2.5. In general, we may restrict to odd n by using Gauss-Bonnet formula. Using orthogonal join, an affirmative example for n = 3 would take care of all n.

In the present form, the rational simplex problem arose recently in the work of Cheeger-Simons [19]. In some sense, the rational simplex problem can be viewed as a special (as well as primary) case of the Archimedean volume problem.

For the Euclidean case to be considered in the next chapter, $\mathcal{P}S/\mathcal{C}S$ can be replaced by $\mathbb{Q}\otimes(\mathcal{P}S/\mathcal{C}S)$. This latter is a commutative, evenly graded, connected Hopf algebra over $\mathbb{Q}$. As an algebra over $\mathbb{Q}$, it is isomorphic to the symmetric algebra over $\mathbb{Q}$ based on the vector space $\mathbb{Q}\otimes\mathcal{P}S/\mathcal{P}S_2$. This is reminiscent of the rational cohomology algebras of classifying spaces of compact, connected, Lie groups. The next problem is a vague speculation based on this crude analogy.

Problem S1. Is there a natural classifying space with cohomology or homology related to $\mathcal{P}S/\mathcal{C}S$? If so, what does it classify?

In a similar speculative vein, commutative Hopf algebras remind us of affine algebraic groups and their generalizations, see Hochschild [38]. The next few problems are in this direction.

Problem S2. Is there an appropriate generalization of affine algebraic groups so that $\mathcal{P}S/\mathcal{C}S$ or $\mathbb{Q}\otimes\mathcal{P}S/\mathcal{C}S$ is the Hopf algebra associated to such objects?

Problem S3. What is the meaning of the higher cohomology groups of the form $H^i(\mathcal{P}S/\mathcal{C}S, \mathcal{P}S)$, $i > 0$? Here, $\mathcal{P}S$ is a right comodule for the Hopf algebra $\mathcal{P}S/\mathcal{C}S$.

Problem S4. Determine the space of primitive elements in the Hopf algebra $\mathcal{P}S/\mathcal{C}S$. More precisely, geometrically characterize the spherical polyhedra A in $S(k^{2i})$ so that $\overline{S\Psi}^{(2j)}\overline{[A]} = 0$ for $0 < j < i$.

7 Euclidean scissors congruence

We begin the present chapter with a modification of the foundation (due to Hadwiger [31]) of the scissors congruence problem in Euclidean spaces of arbitrary dimension. The principal part of the modification consists of the incorporation of the results from the translational as well as the spherical case. In this respect, it is an elaboration of our preliminary manuscript [60]. We then proceed to sketch a proof of Sydler's beautiful theorem showing the completeness of volume and codimensional 2 Dehn invariant in 3-dimensional Euclidean space. Here, we follow the simpler approach due to Jessen [40]. By using results from the translational case, Sydler's theorem extends easily to dimension 4. This extension is due to Jessen [41]. We conclude this chapter with a number of open problems having something to do with the scissors congruence problem in dimensions higher than 4.

1. DOUBLY GRADED ALGEBRA STRUCTURE.

Throughout this chapter, k will denote an (Archimedean) ordered, square root closed field. The conventions of the preceding chapters will remain in force. For convenience, we repeat some of the notations. k^n denotes the k-vector space of column vectors and is given the usual positive definite inner product. The group of motions is the group $E(n;k)$ of all rigid motions so that $E(n;k)$ is the semidirect product of the normal subgroup $T(n;k)$ of all translations by the subgroup $O(n;k)$ of all orthogonal transformations. The direct limits are denoted by k^∞, $E(\infty;k)$, $T(\infty;k)$, $O(\infty;k)$. We often write $E(k)$, $T(k)$ and $O(k)$ for the various groups. The n-simplices are the affine n-simplices of the form $x_0+/x_1/\ldots/x_n/$. We therefore have scissors congruence data. $\mathcal{P}(k^n, E(n;k))$ is abbreviated to $\mathcal{P}E^n$. The square root closure assumption on k shows that $\mathcal{P}E^n$ is unchanged if we allow the n-simplices to range over k^∞ and enlarge $E(n;k)$ to $E(\infty;k)$.

From Chapter 2, $\mathcal{P}E^n$ admits the Minkowski filtration:

$$\mathcal{P}E^n = \mathcal{P}E^n_1 \supset \ldots \supset \mathcal{P}E^n_n \supset \mathcal{P}E^n_{n+1} = \ldots = 0.$$

Using the weights, we have the direct sum decomposition:

$$\mathcal{P}E^n = \coprod_i \mathcal{G}E^n_i, \text{ where } \mathcal{G}E^n_i \cong \mathcal{P}E^n_i/\mathcal{P}E^n_{i+1} \text{ in a canonical way.}$$

Because of the presence of -Id, we have:

$$\mathcal{G}E^n_i = 0 \text{ for } n > 0 \text{ unless } 1 \leq i \leq n \text{ and } n\text{-}i \text{ is even.}$$

For $n > 0$, $\mathcal{G}E^n_i$ and $\mathcal{P}E^n$ are k-vector spaces and the weight space decomposition is a k-vector space decomposition. On $\mathcal{P}E^n$, the k-vector space structure is quite complicated to describe. However, the description is simpler on $\mathcal{G}E^n_i$ by using the canonical isomorphism with $\mathcal{P}E^n_i/\mathcal{P}E^n_{i+1}$. To be more precise, let $A_1\#\ldots\#A_i$ be any basic i-fold cylinder in k^n (it does not have to be orthogonal). If $\lambda \in k^+$, then we have:

$$\lambda[A_1\#\ldots\#A_i] \equiv [A_1\#\ldots\#(\lambda \mathrm{o} A_j)\#\ldots\#A_i] \bmod \mathcal{P}E^n_{i+1}, \quad 1 \leq j \leq i.$$

As a consequence, we have:

$$\lambda^i[A_1\#\ldots\#A_i] \equiv [\lambda \mathrm{o}(A_1\#\ldots\#A_i)] \bmod \mathcal{P}E^n_{i+1}, \quad \lambda \in k^+.$$

The action can be extended to $\mathcal{P}E^n$ through the weight space decomposition and the canonical isomorphisms. The following field theoretic analogue of the Waring problem is now relevant:

<u>Proposition 1.1</u>. Let $n \in \mathbb{Z}^+$. Let F be any field so that the prime field F_0 has cardinality greater than n. Let F_n be the additive subgroup of F generated by all the n-th powers of elements in F. Then $F_n = F$.

<u>Proof</u>. We first show that F_n is always a subfield. By definition, F_n is an additive subgroup and contains 1. Since the generating set is closed under product, the distributive law shows that F_n is a subring. If d is a nonzero element of F_n, then $d^{-1} = d^{n-1} \cdot (d^{-1})^n \in F_n$ because F_n is closed under product. Thus F_n is a subfield; in particular, F_n contains F_0. We next consider the n+1 polynomials $(X+i)^n$ of degree n in $F_0[X]$, $0 \leq i \leq n$. From the binomial theorem, these are F_0-linear transforms of $\binom{n}{i}X^i$, $0 \leq i \leq n$,

where the transforming matrix is just the $(n+1) \times (n+1)$ van der Monde matrix evaluated at $0, 1, \ldots, n$. From the assumption on F_0, $\binom{n}{i} \neq 0$ in F_0 and $0, \ldots, n$ are distinct elements of F_0. As a consequence, X^i, $0 \leq i \leq n$, are F_0-linear combinations of $(X+i)^n$, $0 \leq i \leq n$. Specializing X to $\alpha \in F$, we see that $\alpha \in F_0[F_n] \subset F_n$. Q.E.D.

Remark 1.2. When $n = 2$, the preceding proposition is wellknown. In particular, we can solve $X^2 - Y^2 = \alpha$ by solving $X+Y = 1$ and $X-Y = \alpha$ simultaneously. The preceding proposition can be made effective in the sense that each α can be represented as $\Sigma_i(\beta_i^n - \gamma_i^n)$, $1 \leq i \leq c_n$, where c_n is an effectively computable constant depending only on n.

Corollary 1.3. Let $n \in \mathbb{Z}^+$. Let F be any field so that the prime field F_0 of F has cardinality greater than n. Let $f: V \longrightarrow W$ be any map between F-vector spaces V and W. Then $f \in \mathrm{Hom}_F(V, W)$ if and only if:

(a) f is additive; and

(b) $f(\alpha^n \cdot v) = \alpha^n \cdot f(v)$ holds for all $\alpha \in F$ and $v \in V$.

Proof. The necessity is obvious. The sufficiency follows from Proposition 1.1. Q.E.D.

Let $\mathcal{P}E^+ = \coprod_{n>0} \mathcal{P}E^n$. We can define an associative, commutative, product # on $\mathcal{P}E^+$ through the orthogonal Minkowski sum. More precisely, if P, Q are p-, q-simplices in k^∞, then we can find $\sigma \in E(\infty;k)$ so that $\mathrm{Lsp}(P)$ and $\mathrm{Lsp}(\sigma Q)$ are orthogonal. We then define:

$$[P] \# [Q] = [P \# \sigma Q], \ \sigma \in E(\infty;k), \ <\mathrm{Lsp}(P), \mathrm{Lsp}(\sigma Q)> = 0.$$

As in the spherical case, this product is well defined and associativity commutativity are immediate.

Instead of defining $\mathcal{P}E^0$ to be $\mathbb{Z} \cdot [/./]$ where $[/./]$ denotes the class of a 0-simplex, we define $\mathcal{P}E^0$ to be $k \cdot [/./] \cong k$. Under Minkowski sum, the class $[/./]$ plays the role of the identity element. Using the k-vector space structure, we define:

$$\alpha[/./] \# [P] = \alpha[P] = [P] \# \alpha[/./]$$

There is a minor point involved in this definition. The k-vector space structure was defined on $\mathcal{P}(k^n, T(n;k))$. It is passed onto $\mathcal{P}E^n$. More precisely, $\mathcal{P}E^n$ is a quotient group of $\mathcal{P}^n = \mathcal{P}(k^n, T(n;k))$ through the introduction of the relations:

$$[\sigma P] - [P] = 0, \ \sigma \in O(n;k) \text{ and } P \text{ is a random } n\text{-simplex.}$$

In view of the complicated nature of the k-vector space structure on $\mathcal{P}^n$, it is not immediately evident that the subgroup generated by these relations is actually a k-subspace. In order to avoid the use of dual spaces, we note that the subgroup generated by the relations is stable under scaling by k^+. Since $O(n;k)$ commutes with the scaling action by k^+, the weight decomposition is respected by $O(n;k)$ and the relation subgroup is a graded subgroup. The relation subgroup in $\mathcal{G}^n_i$ is a k-vector subspace by virtue of Proposition 1.1. Specifically, it is an additive subgroup which is closed under scaling corresponding to scalar multiplication by λ^i, $\lambda \in k^+$, $1 \leq i \leq n$. As a consequence, the quotient group $\mathcal{G}E^n_i$ is naturally a k-vector space and our definition makes good sense. We now have:

<u>Theorem 1.4.</u> Let k be an ordered, square root closed field. With the preceding notations and definitions, let $\mathcal{P}E = \coprod_{n \geq 0} \mathcal{P}E^n$. $\mathcal{P}E$ is then a commutative, augmented, doubly graded (by $\mathcal{G}E^n_i$) k-algebra. The space of indecomposable elements, $\mathcal{P}E^+/\mathcal{P}E^+ \cdot \mathcal{P}E^+$ is naturally isomorphic to the direct sum: $\coprod_{i \geq 0} \mathcal{G}E^{2i+1}_1$. We have:

$$\dim_k \mathcal{G}E^{2i+1}_1 = 1 \text{ or } |k| \text{ according to } i = 0 \text{ or } > 0.$$

<u>Proof.</u> For the assertions on the algebra structure, we only need to show the k-bilinearity of the product #. By distributive law, we only need to consider it on the level of the graded components:

$$\mathcal{G}E^m_i \times \mathcal{G}E^n_j \xrightarrow{\#} \mathcal{G}E^{m+n}_{i+j}$$

From the definition, we may restrict our attention to the case where m and n are both positive. By Theorem 3.1.1, scalar multiplication by λ^{m+n}, $\lambda \in k^+$, on either the first, or the second factor on the left side becomes scaling by λ on the right hand side. The latter amounts to scalar multiplication by λ^{m+n} on the right hand side. The k-bilinearity now follows from Corollary 1.3. The assertion on the space of indecomposable elements is a consequence of the canonical isomorphism: $\mathcal{G}E_i^n \cong \mathcal{P}E_i^n/\mathcal{P}E_{i+1}^n$ together with the vanishing of $\mathcal{G}E_i^n$ when n-i is odd, see Proposition 2.5.5. The assertions on dimensions are proved in Proposition 2.5 of the Appendix.

2. DEHN INVARIANTS.

The Dehn invariants can now be defined in the same spirit as in the spherical case. Historically, this is backward. The invariants were first introduced by Dehn for Euclidean 3-space to settle Hilbert's third problem. According to Debrunner [28], the tensor product formulation can already be found in Nicolletti [56] in 1915. According to MacLane [50; p. 172], the concept of tensor product of abelian groups was defined first by Whitney [73] in 1938. Much of the present formulation originated in private letters from Jessen and Thorup.

Let $A = \text{ccl}\{x_0, \ldots, x_n\}$ denote an (affine) n-simplex in k^∞. For each nonempty subset I of $\{0, \ldots, n\}$, let A(I) denote the affine face of A spanned by the vertices x_i, $i \in I$. Randomly select z_I on A(I); for example, the barycenter of A(I) would do. For each $j \in I^c = \{0, \ldots, n\} - I$, let u_j be the unit vector along the normal component of $x_j - z_I$ to Lsp(A(I)). It is then evident that u_j does not depend on the choice of z_I and we can vary z_I over Asp(A(I)). Let $\theta_A(A(I))$ denote the spherical $(n-|I|)$-simplex $\text{sccl}\{u_j | j \in I^c\}$. As in the spherical case, $\theta_A(A(I))$ is called the <u>interior angle</u> at the face A(I) of A. We note that $[A(I)] \in \mathcal{P}E^{|I|-1}$ and $\overline{[\theta_A(A(I))]} \in (\mathcal{P}S/\mathcal{C}S)^{n+1-|I|}$. The <u>total Euclidean Dehn invariant</u> $E\Psi$ of $[A]$ is defined by the formula:

$$E\Psi[A] = \Sigma_I [A(I)] \otimes \overline{[\theta_A(A(I))]} \in \mathcal{P}E \otimes (\mathcal{P}S/\mathcal{C}S),$$

I ranges over all nonempty subsets of $\{0, \ldots, n\}$.

Since $(\mathcal{P}S/\mathcal{C}S)^{2i+1} = 0$, we can restrict to nonempty subsets I with $n-|I|$ even. When $I = \{0, \ldots, n\}$, $A(I) = A$ and $\overline{[\theta_A(A(I))]} = \overline{[\phi]}$ is the unit element of $\mathcal{P}S/\mathcal{C}S$. As in the spherical case, $E\Psi$ is additive with respect to simple subdivision and extends to an additive homomorphism:

$$E\Psi : \mathcal{P}E^{+} \longrightarrow \mathcal{P}E \otimes (\mathcal{P}S/\mathcal{C}S)$$

This is extended to $\mathcal{P}E$ additively by sending $\alpha[/./]$ onto $\alpha[/./]\otimes\overline{[\phi]}$, $\alpha \in k$. The usual associativity formula for tensor products implies that we have an additive map:

$$E\Psi: \mathcal{P}E \longrightarrow \mathcal{P}E\otimes(\mathcal{P}S/\mathcal{C}S) \cong \mathcal{P}E\otimes_k(k\otimes(\mathcal{P}S/\mathcal{C}S)) \qquad (2.1)$$

In view of Theorem 1.4, it is convenient to view $k\otimes(\mathcal{P}S/\mathcal{C}S)$ as a commutative, augmented, bigraded, k-algebra. The bigrading is defined by:

$$(\mathcal{P}S/\mathcal{C}S)^n_i = (\mathcal{P}S/\mathcal{C}S)^n \text{ for } i = 0 \text{ and } = 0 \text{ for } i \neq 0.$$

Since the lower grading refers to weight and angles are invariant under the scaling action by k^+, the bigrading on $\mathcal{P}S/\mathcal{C}S$ is quite natural. However, the finer Minkowski filtration on $\mathcal{P}S^n$ is not reflected by this bigrading. As in Proposition 6.3.7, $k\otimes(\mathcal{P}S/\mathcal{C}S)$ is actually a commutative, evenly graded, connected, Hopf algebra over k. According to the theorem of Hopf-Leray, $k\otimes(\mathcal{P}S/\mathcal{C}S)$ is isomorphic, as a k-algebra only, to the symmetric algebra based on the k-vector space of indecomposable elements: $\coprod_{i>0} k\otimes(\mathcal{P}S^{2i}/\mathcal{P}S^{2i}_2)$. We note that $k\otimes(\mathcal{P}S/\mathcal{C}S)$ loses track of the torsion ideal of $\mathcal{P}S/\mathcal{C}S$. Since $\mathcal{P}E$ is free of additive torsion, this loss has no effect on the strength of the Dehn invariant $E\Psi$. The isomorphism in (2.1) is used to view $\mathcal{P}E\otimes(\mathcal{P}S/\mathcal{C}S)$ as a tensor product over k of two commutative, augmented, bigraded, k-algebra. As such, $\mathcal{P}E\otimes(\mathcal{P}S/\mathcal{C}S)$ is bigraded with:

$$(\mathcal{P}E\otimes(\mathcal{P}S/\mathcal{C}S))^n_i = \coprod_{0\leq j\leq n} \mathcal{G}E^j_i \otimes (\mathcal{P}S/\mathcal{C}S)^{n-j}.$$

From the definition of $E\Psi$, it is evident that we have compatibility with respect to scaling by k^+ in the sense that:

$$E\Psi[\lambda \circ A] = \Sigma_I [\lambda \circ A(I)] \otimes \overline{[\theta_A(A(I))]}$$

As a consequence, $E\Psi$ preserves the double grading. Concentrating on the various weight components and using Proposition 1.1, $E\Psi$ can be seen to be k-linear. In a manner completely similar to the proof of Proposition 6.3.17, we have the following result:

Proposition 2.2. Let k be an ordered, square root closed field. The total Euclidean Dehn invariant $E\Psi$ is a homomorphism of commutative, bigraded, augmented, k-algebras. With $k \otimes (\mathcal{P}S/\mathcal{C}S)$ viewed as a commutative, evenly graded, connected, Hopf algebra over k, $E\Psi$ is also the structure map turning $\mathcal{P}E$ into a graded, (right) comodule for $k \otimes (\mathcal{P}S/\mathcal{C}S)$. As such, the subgroups $\coprod_{0 \leq j \leq n} \mathcal{P}E^j$, $n \geq 0$, $\mathcal{P}E^{even}$, $\mathcal{P}E^{odd}$ are subcomodules.

As in the spherical case, we can define the graded volume map on $\mathcal{P}E$:

$$\text{gr. vol.} : \mathcal{P}E \longrightarrow k[T].$$

If A is an n-simplex, then gr. vol. $[A] = \text{vol}_n(A)T^n$ with $\text{vol}_n(A)$ denoting the absolute n-dimensional volume of A. Unlike the spherical case, $\text{vol}_n(A)$ is defined through determinants and made sense as long as k is ordered. The positive definite inner product can be used to normalize the definition. As a consequence, gr. vol. is a homomorphism of commutative, bigraded, augmented, k-algebras. Here T^i is assigned the bidegree (i, i). We can now define the classical total Euclidean Dehn invariant $E\Phi$:

$$E\Phi = (\text{gr. vol.} \otimes \overline{\text{Id}}) \circ E\Psi : \mathcal{P}E \longrightarrow k[T] \otimes (\mathcal{P}S/\mathcal{C}S).$$

In view of the earlier observation, $k[T] \otimes (\mathcal{P}S/\mathcal{C}S)$ is also isomorphic, as a k-algebra only, to a symmetric algebra over k. Indeed, we have:

> $k[T] \otimes (\mathcal{P}S/\mathcal{C}S)$ is isomorphic, as a k-algebra, to a symmetric algebra over k based on $kT \coprod (\coprod_{i>0} k \otimes (\mathcal{P}S^{2i}/\mathcal{P}S^{2i}_2))$.

It follows that the injectivity of $E\Phi$ would force $\mathcal{P}E$ to be an integral domain. This is one of the major open problems.

As in the spherical case, $E\Psi$ and $E\Phi$ can be broken down to their codimensional i components:

$$E\Psi = \Sigma_{i \geq 0} E\Psi^{(i)}, \text{ and } E\Phi = \Sigma_{i \geq 0} E\Phi^{(i)}.$$

$E\Psi^{(0)}$ identifies $\mathcal{P}E$ with $\mathcal{P}E \otimes \overline{[\emptyset]}$ while $E\Phi^{(0)}$ is essentially the graded volume map. In general, we have:

$$E\Psi^{(i)} : \mathcal{G}E^n_j \longrightarrow \mathcal{G}E^{n-i}_j \otimes (\mathcal{P}S/\mathcal{C}S)^i.$$

It follows that:

$E\Psi^{(i)}$ is nonzero on $\mathcal{G}E^n_j$ only when $i \geq 0$ is even and $1 \leq j \leq n-i$ with $n-j$ even or with $i = n = j = 0$.

We note that the case $i = n > 0$ contains the assertion that the sum of the vertex angles of an n-simplex has its class in $\mathcal{C}S$ when $n > 0$. In the same manner, we have:

$E\Phi^{(i)}: \mathcal{G}E^n_j \longrightarrow kT^{n-i} \otimes (\mathcal{P}S/\mathcal{C}S)^i$; and $E\Phi^{(i)}$ is nonzero on $\mathcal{G}E^n_j$ only when $i \geq 0$ is even and $j = n-i$.

In a slightly different form, the following result was communicated to us in a private letter from Jessen and Thorup:

Proposition 2.3. Let k be an ordered, square root closed field. With the preceding notations and definitions, the following statements are all equivalent:

(a) $E\Phi$ is injective.

(b) For each $n \geq 0$, the maps $E\Phi^{(2i)}$, $0 \leq i < n/2$, separate the points of the k-vector space $\mathcal{P}E^n$.

(c) $H^0(k\otimes(\mathcal{P}S/\mathcal{C}S), \mathcal{P}E) = \coprod_{n\geq 0} \mathcal{G}E^n_n$.

(d) For each $n \geq 0$, the maps $E\Psi^{(2i)}$, $0 < i < n/2$, together with the graded volume map $E\Phi^{(0)}$ separate the points of the k-vector space $\mathcal{P}E^n$.

Proof. The equivalence of (a) and (b) is a consequence of the discussion preceding the proposition. From the definition, $H^0(k\otimes(\mathcal{P}S/\mathcal{C}S), \mathcal{P}E)$ is the graded k-subspace $\bigcap_{i>0} \ker E\Psi^{(2i)}$. From the multiplicative property of $E\Psi$, it is immediate that:

$$\coprod_{n\geq 0} \mathcal{G}E^n_n \subset \bigcap_{i>0} \ker E\Psi^{(2i)}.$$

From the weight decomposition, it is evident that:

$$\ker E\Phi^{(0)} = \coprod_{i,n>0} \mathcal{G}E^n_{n-2i}.$$

The equivalence of (c) and (d) is therefore clear.

Assume (b) holds. Let $x \in \mathcal{P}E^n$ with $E\Phi^{(0)}(x) = 0 = E\Psi^{(2i)}(x)$, $0 < i < n/2$. Since $E\Phi^{(2i)}(x) = (\text{gr. vol.} \otimes \text{Id})\{E\Psi^{(2i)}(x)\}$, (d) follows from (b).

Conversely, assume (d) and let $x \in \mathcal{P}E^n$. (d) shows that all the information about x can be recovered from $E\Phi^{(0)}(x) \in \mathcal{G}E^n_n \otimes \overline{[\phi]}$ and from $E\Psi^{(2i)}(x) \in \mathcal{G}E^{n-2i} \otimes (\mathcal{P}S/\mathcal{C}S)^{2i}$, $0 < i < n/2$. Weight space decomposition shows that $E\Phi^{(0)}(x)$ carries exactly all the information about the component of x in $\mathcal{G}E^n_n$ while $E\Psi^{(2i)}(x)$, $0 < i < n/2$, carries exactly all the information about the components of x in $\mathcal{G}E^n_j$, $j < n$. Since $n - 2i < n$, (b) can now be verified through induction on n. Q.E.D.

As mentioned in the spherical case, we can iterate the various Dehn invariants through composition to obtain other invariants. In particular, we have the <u>total classical, Euclidean, Dehn-Hadwiger invariant</u>:

$$(\text{Id} \otimes \xi_2) \circ E\Phi : \mathcal{P}E \longrightarrow k[T] \otimes_k \text{Sh}_k(k \otimes (\mathcal{P}S^2/\mathcal{P}S^2_2)) \tag{2.4}$$

It is a homomorphism of commutative, bigraded, augmented, k-algebra. T is given the bidegree (1, 1) while $\mathcal{P}S^2/\mathcal{P}S^2_2$ is given the bidegree (2, 0). It is not known how much (if any) information is lost when $E\Phi$ is replaced by $(\text{Id} \otimes \xi_2) \circ E\Phi$. In the appendix, this invariant is used to determine the size of $\mathcal{G}E^n_j$.

3. THEOREM OF SYDLER.

We now come to one of the major results on scissors congruence.

<u>Theorem 3.1</u>. Let k be an ordered, square root closed field. The Dehn invariants $E\Phi^{(0)}$ and $E\Phi^{(2)}$ separate the points of $\mathcal{P}E^n$ when $n \leq 4$.

When $n \leq 2$, the graded volume $E\Phi^{(0)}$ already separates the points of $\mathcal{P}E^n$. When $n = 4$, $\mathcal{P}E^4 = \mathcal{G}E^4_4 \coprod \mathcal{G}E^4_2$. $E\Phi^{(0)}$ is injective on $\mathcal{G}E^4_4$ and 0 on $\mathcal{G}E^4_2$. From Theorem 3.1.1, $\mathcal{G}E^4_2 = [/e_1/] \# \mathcal{G}E^3_1$. The multiplicative property of $E\Phi$ shows that $E\Phi^{(2)}$ is injective on $\mathcal{G}E^4_2$ if and only if it is injective on $\mathcal{G}E^3_1$. This argument originated in Jessen [41]. We are therefore reduced to the case n = 3. The result is due to Sydler [67]; it appears to be the culmination of a sequence of papers written during 1943-1965. We will consider the simpler treatment due to Jessen [40].

Jessen's treatment assumed that $k = \mathbb{R}$. A proof analysis shows that the result is valid under the weaker assumptions as stated. In essence, the geometric constructions were of ruler-compass nature. The ordering is needed to make sense out of the concept of subdivision.

To begin the proof analysis, we consider the concept of angles. In the case of $k = \mathbb{R}$, angles in the plane are measured in radians. In Jessen's proof, the basic angles are acute (though their sum need not be acute). In terms of the group $\mathcal{P}S^2/\mathcal{P}S^2_2$, acute angles are in bijective correspondence with the nonzero elements of $\mathcal{P}S^2/\mathcal{P}S^2_2$. In terms of the coordinatization of $SO(2;k)$, an acute angle θ can be identified with $(c(\theta), s(\theta))$, where:

$$c(\theta)^2 + s(\theta)^2 = 1, \; c(\theta), \; s(\theta) \in k^+ \text{ so that } 0 < c(\theta), \; s(\theta) < 1.$$

Since k is ordered and square root closed, the nonzero elements of $\mathcal{P}S^2/\mathcal{P}S^2_2$ are in bijective correspondence with $]0,1[$ through the function s (or c). This observation bypasses the need of the Archimedean assumption on k as far as angles go. The usual trignometric functions are rational functions of c and s, i.e., elements of the function field $k(s,c)$ where $s^2+c^2 = 1$. We note that $\mathcal{P}S^2_2 \cong \mathbb{Z}$ and our normalization is such that $\theta = 1$ corresponds to the angle $\pi/2$ in radian measure.

We now describe the general outline of the proof of Sydler's theorem. We already know that $\mathcal{P}E^3 = \mathcal{G}E^3_3 \coprod \mathcal{G}E^3_1$ and that the (graded) volume map $E\Phi^{(0)}$ defines an isomorphism between $\mathcal{G}E^3_3$ (more generally, $\mathcal{G}E^n_n$) and k and vanishes on $\mathcal{G}E^3_1$ (more generally, $\mathcal{G}E^n_j$ with $j < n$). Similarly, the Dehn invariant $E\Psi^{(2)}$ vanishes on $\mathcal{G}E^3_3$ (more generally, $\mathcal{G}E^n_j$ with $j \neq n-2$) and defines a homomorphism (of k-vector spaces):

$$E\Psi^{(2)} : \mathcal{G}E^3_1 \cong \mathcal{P}E^3/\mathcal{P}E^3_3 \longrightarrow \mathcal{P}E^1 \otimes (\mathcal{P}S^2/\mathcal{P}S^2_2) \qquad (3.2)$$

The volume map defines an isomorphism between $\mathcal{P}E^1$ and k and defines an injective map from $\mathcal{P}S^2$ to k carrying $\mathcal{P}S^2_2$ onto $\mathbb{Z} \subset k$. We can therefore identify the right hand side of (3.2) with a subspace of $k \otimes (k/\mathbb{Z})$ with the understanding that the vector space structure comes from the first factor. Composing $E\Psi^{(2)}$ with the various inclusions and injections and dropping the grading, we have:

$$E\Phi^{(2)} : \mathcal{P}E^3/\mathcal{P}E^3_3 \longrightarrow k\otimes(k/\mathbb{Z}) \quad (3.3)$$

Sydler's theorem is equivalent with the injectivity of $E\Phi^{(2)}$ in (3.3). Using duality theory of vector spaces, this is then equivalent with the surjectivity of $E\Phi^{(2)*}$ on dual spaces:

$$\begin{aligned} E\Phi^{(2)*} &: \mathrm{Hom}_k(k\otimes(k/\mathbb{Z}), k) \cong \mathrm{Hom}(k/\mathbb{Z}, k) \\ &\longrightarrow \mathrm{Hom}_k(\mathcal{P}E^3/\mathcal{P}E^3_3, k) \cong J^3_1(k). \end{aligned} \quad (3.4)$$

In (3.4), the target group k can be replaced by any nonzero k-vector space; for example, we can use $\mathcal{P}E^3/\mathcal{P}E^3_3 \cong \mathcal{G}E^3_1$. In other words, Sydler's theorem is equivalent with the following assertion:

> Let $\tau : \mathcal{P}E^3/\mathcal{P}E^3_3 \longrightarrow Y$ be a k-linear map into a nonzero k-vector space Y. Then there is an additive map:
>
> $$\begin{aligned} &\varphi_\tau : k \longrightarrow Y, \ \varphi(1) = 0 \text{ and} \\ &\tau[A] = \Sigma_i \ell_i \varphi_\tau(\alpha_i) \end{aligned} \quad (3.5)$$
>
> where A is any polyhedron in k^3 with normalized interior dihedral angle α_i (measured in k via volume) at the edge ℓ_i (measured in k via volume).

Since τ is k-linear in (3.5), the polyhedron A can be restricted to any set of k-vector space generators of $\mathcal{P}E^3$. We already know that $\mathcal{P}E^3$ can be generated by the 3-parameter family of orthogonal 3-simplices of the form: $[/ae_1/be_2/ce_3/]$, a, b, $c \in k^+$. This is a generation as an additive group. As a k-vector space, the k^+ scaling action cuts it down to a 2-parameter family. This amounts to the imposition of a relation among a, b and c. The discovery of Sydler was that the following special 2-parameter family of orthogonal 2-simplices T(a, b) or $T(\alpha, \beta)$ enjoys a number of properties that ultimately led to a proof of (3.5):

$$\begin{aligned} &T(a,b) = T(\alpha,\beta) = /\mathrm{ctg}(\alpha)e_1/\mathrm{ctg}(\alpha)\mathrm{ctg}(\beta)e_2/\mathrm{ctg}(\beta)e_3/ \\ &\text{where } 0 < a,\ b,\ \alpha,\ \beta < 1;\ s(\alpha)^2 = a,\ s(\beta)^2 = b; \\ &\mathrm{ctg}(\theta) = c(\theta)/s(\theta),\ c(\theta) = (1-s(\theta))^{1/2} \in k^+,\ \theta = \alpha,\ \beta. \end{aligned} \quad (3.6)$$

Up to a difference in the normalization of angles, ctg is just the cotangent function. Similarly, tg will denote the tangent function. Since k is ordered and square root closed, the equation: $s(\alpha)^2 = a$, defines a self bijection of $]0,1[$. We already noted that the acute angles (the nonzero elements of $\mathcal{P}S^2/\mathcal{P}S^2_2$) are in bijective correspondence with the points of $]0,1[$. Under multiplication, $]0,1[$ is an ordered, commutative semigroup without identity and has k^+ as its universal group. The sine square bijection allows us to transport the structure to acute angles. More precisely, we define the $\star$ product between acute angles α and β as follows:

$$s(\alpha\star\beta)^2 = s(\alpha)^2 s(\beta)^2, \text{ equivalently, } s(\alpha\star\beta) = s(\alpha)s(\beta)$$

where $0 < \alpha,\ \beta,\ \alpha\star\beta < 1$.

It is now an easy calculation to show that:

$$E\Phi^{(2)}[T(a,b)] = ctg(\alpha)\otimes\overline{\alpha} + ctg(\beta)\otimes\overline{\beta} - ctg(\alpha\star\beta)\otimes\overline{(\alpha\star\beta)};$$

$$E\Phi^{(0)}[T(a,b)] = v(a)+v(b)-v(ab), \text{ where } v(c) = (c-1)/6c.$$

In these formulas, we have dropped the grading and applied the volume maps (they are injective in these ranges). If we temporarily view $k\otimes(k/\mathbb{Z})$ and k as k^+-trivial modules, then the preceding formulas indicate that we have 2-coundaries for $]0,1[$ under the product in k. It is now immediate:

$[T(b,c)] + [T(a,bc)]$ and $[T(ab,c)]+[T(a,b)]$
have the same Dehn invariants $E\Phi^{(2)}$ and $E\Phi^{(0)}$,
$0 < a,\ b,\ c < 1$ are arbitrary in k.

The following fundamental lemma of Sydler is then a special case of his main theorem:

Lemma 3.7. Let k be an ordered, square root closed field. Let a, b, c be arbitrary in k with $0 < a,\ b,\ c < 1$. Then:

$$[T(b,c)] + [T(a,bc)] = [T(ab,c)] + [T(a,b)] \text{ in } \mathcal{P}E^3.$$

We note that T(a, b) and T(b, a) are O(3;k)-congruent. Our a, b and c correspond to b, a and c in Jessen [40; Lemma 1, p. 247]. A careful analysis of Jessen's presentation shows that all the geometric constructions are of ruler-compass variety so that it holds under the hypotheses as stated. The proof of this lemma has already been reproduced in Boltianskii [11;144-147], we will refrain from further reproduction. We note that the equality of the volumes allows us to work in $\mathcal{P}E^3/\mathcal{P}E^3_3$. The stable scissors congruence can be replaced by scissors congruence when k is Archimedean (this was due to Sydler for $\mathbb{R}^3$ in earlier works). (3.5) now becomes:

If τ is defined as in (3.5), then there is an additive map φ_τ: k ⟶ Y, $\varphi_\tau(1) = 0$ and $\tau[T(a,b)] = \text{ctg}(\alpha)\varphi_\tau(\alpha)+\text{ctg}(\beta)\varphi_\tau(\beta) - \text{ctg}(\alpha\star\beta)\varphi_\tau(\alpha\star\beta)$, $0 < a, b < 1$, $a = s(\alpha)^2$, $b = s(\beta)^2$. (3.8)

If $Y = \mathcal{G}E^3_1$ and $\tau = \text{Id}$, then (3.8) amounts to a suitable interior disjoint union representation of T(a, b) modulo parallelopipeds in terms of polyhedra determined by the function φ_τ. Aside from two easier geometric lemmas (also found in Sydler's work), Jessen showed that the existence of φ_τ involves only arguments of homological nature (though fairly delicate). We will sketch these. Part of the homological algebra amounts to extending the well known facts on cohomology of groups and rings to suitable commutative semigroups and semirings. This is done in Jessen, Karpf and Thorup [43]. To some extent, part of these are already contained in Cartan-Eilenberg [15]. For the purpose at hand, [43] is more appropriate. From Lemma 3.7, we can find a function f:]0, 1[⟶ Y so that:

$$\tau[T(a,b)] = f(a) + f(b) - f(ab), \quad 0 < a, b < 1 \tag{3.9}$$

The point is that the left hand side of (3.9) is a symmetric 2-cocycle of]0, 1[with values in the trivial k^+-module Y. Since Y is divisible, such a 2-cycle must be a coboundary. f is only unique up to the addition of a homomorphism from]0, 1[, · to Y, + . The idea is to straighten out f so that we can eventually find a φ_τ satisfying (3.8). The next result from Sydler's work (see Jessen [40; Lemma 2, p. 249]) is:

Lemma 3.10. Let k be an ordered, square root closed field. Let a, b, c lie in k^+. Then: $[a \circ T(\frac{a+b}{a+b+c}, \frac{a}{a+b})] + [b \circ T(\frac{a+b}{a+b+c}, \frac{b}{a+b})]$ is equal to $[a \circ T(\frac{a+c}{a+b+c}, \frac{c}{a+c})] + [c \circ T(\frac{a+c}{a+b+c}, \frac{c}{a+c})]$ in $\mathcal{P}E^3$.

The complicated appearance of this lemma is quite deceiving. The proof is obtained by cutting the simplex $\mathrm{ccl}\{0, x_1, x_2, x_3\}$ in two different ways. Here $x_1 = (bc)^{1/2} e_1$, $x_2 = (ca)^{1/2} e_2$, $x_3 = (ab)^{1/2} e_3$. The two terms on the left hand side come from cutting by using the plane through 0, x_3 and orthogonal to the opposite edge $\mathrm{ccl}\{x_1, x_2\}$. The two terms on the right hand side come from cutting by using the plane through 0, x_2 and orthogonal to the opposite edge $\mathrm{ccl}\{x_1, x_3\}$. We note that the proof actually shows scissors congruence rather than stable scissors congruence. We note also that $\mathrm{ccl}\{0, x_1, x_2, x_3\}$ is not orthogonal. The recent paper of Jessen [42] discusses the spherical and hyperbolic analogues.

From the function f described in (3.9), we can form the new 2-cochain:

$$G : k^+ \times k^+ \longrightarrow Y \text{ so that:}$$

$$G(a, b) = a f\left(\frac{a}{a+b}\right) + b f\left(\frac{b}{a+b}\right).$$

Lemma 3.10 together with the definition of G immediately yields:

$$G(a, b) = G(b, a); \; G(\lambda a, \lambda b) = \lambda G(a, b)$$

$$G(b, c) + G(a, b+c) = G(a+b, c) + G(a, b)$$

At this point, G can be extended to k and still satisfy the same functional equations, see [43]. It can be viewed as a 2-cocycle defining an exact sequence of k-vector spaces. It is therefore a coboundary. More precisely, we can find a function $g_1 : k^+ \longrightarrow Y$ so that:

$$g_1(ab) = g_1(a) + g_1(b) \text{ and}$$

$$G(a, b) = g_1(a) + g_1(b) - g_1(a+b).$$

Writing g(a) for $g_1(a)/a$, we have a function $g : k^+ \longrightarrow Y$ so that:

$$g(ab) = g(a) + g(b); \text{ and}$$

$$G(a, b) = ag(a) + bg(b) - (a+b)g(a+b)$$

We note that $g(1) = 0$. As a consequence, we have:

If $a, b \in k^+$ with $a+b = 1$, then:

$$af(a) + bf(b) = ag(a) + bg(b).$$

If we define $h = f - g :]0,1[\longrightarrow Y$, then we have:

$$\begin{aligned} &\tau[T(a,b)] = h(a) + h(b) - h(ab); \text{ and} \\ &ah(a) + bh(b) = 0 \text{ if } a+b = 1,\ a,\ b > 0. \end{aligned} \tag{3.11}$$

This is the desired modification of f in the sense that we can define φ_τ by:

$$\begin{aligned} &\varphi_\tau : k \longrightarrow Y,\ \varphi_\tau \text{ is periodic with perior } 1; \\ &\varphi_\tau(1) = 1; \text{ and } \varphi_\tau(\theta) = tg(\theta)\, h(s(\theta)^2),\ 0 < \theta < 1. \end{aligned} \tag{3.12}$$

It follows that:

$$\begin{aligned} &\varphi(\alpha+\beta) = 0 \text{ if } \alpha,\ \beta \in k \text{ with } \alpha+\beta \in \mathbb{Z}; \text{ and} \\ &\tau[T(a,b)] = ctg(\alpha)\varphi_\tau(\alpha) + ctg(\beta)\varphi_\tau(\beta) - ctg(\alpha\star\beta)\,\varphi_\tau(\alpha\star\beta). \end{aligned} \tag{3.13}$$

The proof of (3.9), hence of Theorem 3.1, would be complete as soon as we verify that φ_τ is additive. For this we observe that τ vanishes on $\mathcal{P}E_3^3 = \mathcal{P}E_2^3$. We state the next elementary lemma used by Sydler: (see [40; Lemma 3])

Lemma 3.14. Let k be an ordered, square root closed field. Let α, β, γ be acute angles with sum 2 (equals to half of $S(k^2)$ or π radians when k is Archimedean). Then there is an orthogonal 3-simplex $/x_1/x_2/x_3/$ such that the interior dihedral angles along the edge $x_1+x_2+x_3$ of $/x_1/x_2/x_3/$, $/x_2/x_1/x_3/$ and $/x_2/x_3/x_1/$ are respectively α, β and γ.

We note that $x_1+x_2+x_3$ is an diagonal of the rectangular parallelopiped: $/x_1/\#/x_2/\#/x_3/$. Half of this is $/x_1/\#/x_2/x_3/$ and is the interior disjoint union of $/x_1/x_2/x_3/$, $/x_2/x_1/x_3/$ and $/x_2/x_3/x_1/$. A simple computation with (3.13) using the additivity of τ and its vanishing on $\mathcal{P}E_3^3$ quickly yields:

$$\begin{aligned} &\varphi_\tau(\alpha) + \varphi_\tau(\beta) + \varphi_\tau(\gamma) = 0 \text{ if } 0 < \alpha,\ \beta,\ \gamma < 1 \text{ and} \\ &\alpha + \beta + \gamma = 2. \end{aligned}$$

This together with (3.13) and the periodicity of φ_τ quickly lead to the additivity of φ_τ.

The preceding proof of Theorem 3.1 actually yields a description of the image of $E\Phi^{(2)}$. The result is due to Jessen [40]. In some sense, this is analogous to the syzygy problem for the translational case.

Theorem 3.15. Let k be an ordered, square root closed field. Then the image of $E\Phi^{(2)} : \mathcal{P}E^3/\mathcal{P}E_3^3 \longrightarrow k \otimes (k/\mathbb{Z})$ consists of all the elements of the form: $\Sigma_i \ell_i \otimes \overline{\theta}_i$ such that $\Sigma_i \ell_i d(s(\theta_i))/c(\theta_i) = 0$ where $d : k \longrightarrow k$ ranges over all the absolute derivations (of k over $\mathbb{Q}$).

To see this, we again use duality in vector spaces. Since $E\Phi^{(2)}$ is injective, the study of the image of $E\Phi^{(2)}$ amounts to the study of the maps φ_τ with $\tau = 0$. In view of (3.11) and (3.12), these φ_τ's are defined by:

$$\begin{aligned} &\varphi_\tau : k \longrightarrow k \text{ is additive, } \varphi(1) = 0; \\ &\varphi_\tau(\theta) = tg(\theta)\, h(s(\theta)^2),\ 0 < \theta < 1; \text{ and} \\ &h : \,]0,1[\, \longrightarrow Y \text{ satisfies: } h(ab) = h(a) + h(b) \\ &\text{and } ah(a)+bh(b) = 0 \text{ if } a+b = 1. \end{aligned} \tag{3.16}$$

If $d : k \longrightarrow Y$ is a derivation over $\mathbb{Q}$ so that $d(a+b) = da + db$ and so that $d(ab) = bd(a) + ad(b)$, then $h(a) = d(a)/a$, $0 < a < 1$, satisfies the condition (3.16). Conversely, if h satisfies the condition in (3.16), then $d(a) = ah(a)$ can be extended uniquely to a derivation, see [40 ; p 254] for the details. As a consequence, $tg(\theta)h(s(\theta)^2)$ becomes $2\,d(s(\theta))/c(\theta)$. If we specialize Y to k, then the image of $E\Phi^{(2)}$ in $k \otimes (k/\mathbb{Z})$ is simply the intersection of the kernels associated to the derivations $d: k \longrightarrow k$. This is just the assertion in theorem 3.15. Since every derivation of k is necessarily zero on the algebraic numbers in k over $\mathbb{Q}$, we have the following result of Jessen:

Corollary 3.16. Let k be an ordered, square root closed field. An element of the form $\ell \otimes \overline{\theta}$, $\ell \in k^\times$, lies in the image of $E\Phi^{(2)}$ if and only if $s(\theta)$ is algebraic over $\mathbb{Q}$ (equivalently, $e(\theta) = c(\theta) + \iota\, s(\theta)$ is algebraic over $\mathbb{Q}$).

4. OPEN PROBLEMS.

We begin by rephrasing the scissors congruence problem.

Strong Scissors Congruence Problem. (Euclidean Case.) Let k denote an (Archimedean) ordered, square root closed field.

Algebra version. Find enough reasonable k-algebra homomorphisms (respecting all structures) to separate the points of $\mathcal{P}E$.

Comodule version. Find enough reasonable (linear or semilinear) comodule homomorphisms (respecting all structures) to separate the points of $\mathcal{P}E$ as a (right) comodule over the Hopf algebra $k \otimes (\mathcal{P}S/\mathcal{C}S)$.

The following problem is a special instance of the algebra version:

Dehn Invariant Problems. Let k be an ordered, square root closed field. Find the kernel of $E\Phi$ and of $(Id \otimes \xi_2) \circ E\Phi$. In particular, are they 0 ?

The injectivity of $E\Phi$ or $(Id \otimes \xi_2) \circ E\Phi$ would give an affirmative answer to the following problem:

Integral Domain Problem. Let k be an ordered, square root closed field. Is $\mathcal{P}E$ an integral domain ?

We note that the analogous question for the spherical case had an affirmative answer modulo torsion when k is Archimedean. Propositions 2.2.1 and 2.5.5 together imply that the space of indecomposable elements in $\mathcal{P}E$ is $\coprod_{i \geq 0} \mathcal{G}E_1^{2i+1}$. This suggests the following:

Algebra Structure Problem. Let k be an ordered, square root closed field. Is $\mathcal{P}E^{2i}$ the direct sum of the subspaces $\mathcal{G}E_1^{2j(1)+1} \# \ldots \# \mathcal{G}E_1^{2j(t)+1}$, where $2j(s)+1$ range over all the odd partitions of 2i. Is $\mathcal{P}E$ isomorphic to a symmetric algebra based on the k-vector space of indecomposable elements? Is $\mathcal{P}E$ a Hopf algebra?

If $\mathcal{P}E$ is a Hopf algebra in a natural way, then the algebra structure of $\mathcal{P}E$ is clarified by the theorem of Hopf-Leray. The Dehn invariant problem can then be reduced to the injectivity of the Dehn invariants on the space of indecomposable. We note that our graded structures on $\mathcal{P}E \otimes (\mathcal{P}S/\mathcal{C}S)$ did not take into account of the Minkowski filtration on $\mathcal{P}S$.

Minkowski Filtration Problem. Let k be an ordered, square root closed field. Pull back the Minkowski filtration on $\mathcal{P}S$ to $\mathcal{P}E$ via the Dehn invariant $E\Psi$. Geometrically characterize this induced filtration on $\mathcal{P}E$.

We end this section with a few loose remarks. In dimension 5, the critical part of the scissors congruence problem is the separation of $\mathcal{G}E_1^5$. By using weight, the only classical Dehn invariant of any import is $E\Phi^{(4)}$ with its value in $k \otimes (\mathcal{P}S/\mathcal{C}S)^4$ where k is identified with $\mathcal{P}E^1$. If we compute this Dehn invariant for a general orthogonal 5-simplex, we would get 6 terms. Of these, 3 lie in $k \otimes (\mathcal{P}S_2^4/\mathcal{C}S^4)$, $\mathcal{C}S^4 = \mathcal{P}S_3^4$. In view of the pattern of the proof of Sydler's theorem, this suggests that a two step generalization may be plausible. More precisely, the first step is to work with $k \otimes (\mathcal{P}S^4/\mathcal{P}S_2^4)$. After that, work with $k \otimes (\mathcal{P}S_2^4/\mathcal{P}S_3^4)$. In both of these cases, we should work directly with $\mathcal{P}S^4$ rather than "numerical" invariants. This means that we argue geometrically. In order to have some chance of generalization, we need to develope a manageable "bookkeeping device". It is not clear that our primitive notation for spherical simplices and orthogonal joins is adequate for such a purpose. When we come to dimension 6, a new problem arises. Jessen's extension of Sydler's theorem is based on the simple fact that $\mathcal{P}E^4 = \mathcal{P}E^1 \# \mathcal{P}E^3$. In dimension 6, we have the result:

$$\mathcal{P}E^6 = (\mathcal{P}E^1 \# \mathcal{P}E^5) + (\mathcal{P}E^3 \# \mathcal{P}E^3).$$

Using the multiplicative property of $E\Phi$ together with Sydler's theorem, the important part of $\mathcal{P}E^6$ is $\mathcal{P}E^1 \# \mathcal{P}E^5$. As it stands, we do not have a direct sum decomposition. If we pass to the weight components, we have:

$$\mathcal{G}E_2^6 = (\mathcal{G}E_1^1 \# \mathcal{G}E_1^5) + (\mathcal{G}E_1^3 \# \mathcal{G}E_1^3).$$

If this can be shown to be a direct sum, then it would mean that in dimension 5, we essentially only have to worry about $k \otimes (\mathcal{P}S^4/\mathcal{P}S_2^4)$ as far as the target of $E\Phi^{(2)}$ goes. These are the crude reasons behind some of the open problems. It is apparent some experimentations with special polyhedra or simplices are required. We hope to come back to these.

In addition to these vague remarks, it is irresistible to repeat several cryptic observations.

The first is due to Dennis Sullivan. The target group of the codimension 2 Dehn invariant is $\mathbb{R} \otimes (\mathbb{R}/\mathbb{Z})$ when $k = \mathbb{R}$. This has $K_2(\mathbb{C})^-$ as a quotient, see Sah-Wagoner [61]. What is the connection (if any)? (This was the observation that got us interested in the third problem.)

The second is due to Jessen and Thorup in a letter to us. It is connected with the spherical case. We include it here because it is connected with derivations--which appeared in Jessen's theorem (Theorem 3.15). Again, we assume $k = \mathbb{R}$. Since $\mathbb{R}$ is a field, we have a universal derivation:

$$d : \mathbb{R} \longrightarrow \Omega_{\mathbb{R}/\mathbb{Q}},$$

Here $\Omega_{R/\mathbb{Q}}$ denotes the $\mathbb{R}$-module of all Kähler differentials, see Mumford [55; p. 279]. We note that d factors through $\mathbb{R}/\mathbb{Z}$. If we follow the modified spherical Dehn invariant (see p. 124) $S\Phi^{(2)} - 1 \otimes S\Phi^{(0)}$ by $1 \otimes d$, then by the multiplication map, then a spherical 3-simplex S is mapped onto:

> $\Sigma_i \ell_i d(\theta_i) - d(\mathrm{vol}_3 S)$, where θ_i is the normalized interior dihedral angle of S at the edge of length ℓ_i.

This formula defines an additive homomorphism from $\mathcal{P}S^4/\mathcal{P}S^4_3$ into $\Omega_{\mathbb{R}/\mathbb{Q}}$. It should be noted that $\Omega_{\mathbb{R}/\mathbb{Q}}$ is an $\mathbb{R}$-vector space of dimension equal to the cardinality of $\mathbb{R}$. The kernel of the universal derivation d is just the algebraic closure of $\mathbb{Q}$ in $\mathbb{R}$. The cryptic observation is the comparison of the preceding formula with Schläfli's analytic differential equation:

> $d(\mathrm{vol}_3 S) = \Sigma_i \ell_i d(\theta_i)$, d now denotes the analytic differential with $\mathrm{vol}_3 S$ viewed as an analytic function of the interior dihedral angles.

In some vague sense, these resemble the results of Harris [36].

Finally, Chern observed in a private conversation that Dehn invariants appear to be similar to integrals of mean curvature and quermassintegrale, see Santalo [62; p. 226, formula (13.58)] as well as Hadwiger [31] for more comprehensive studies. All of these suggest the tips of a yet to be worked out "combinatorial differential and/or integral geometry". See also the open problems in the next chapter.

8 Hyperbolic miscellanies

Among the classical geometries, the hyperbolic case of the scissors congruence problem is perhaps the most intriguing--the geometry of angles is spherical while the geometry of a horosphere is Euclidean. The positive result in the hyperbolic case is similar to the spherical case--length and area are complete invariants in dimensions 1 and 2 respectively. For higher dimensions, the Dehn invariants are still necessary conditions for scissors congruence and they lead to a (right) comodule structure on the group $\mathcal{PH}$ with respect to the Hopf algebra $\mathcal{PS}/\mathcal{CS}$. In contrast to the spherical and Euclidean cases, we do not have the analogue of a ring structure, nor do we have a reduction from even to odd dimension. Most of the present chapter concerns itself with open problems as well as some loose ideas. Some of these are only vaguely connected with the scissors congruence problem. A basic reference on the geometry and topology of hyperbolic spaces is the 1977-78 Princeton lecture notes of Thurston [68]. These notes contain a wealth of new results on the geometry and topology of 3-manifolds as well as many other related matters. In addition, we have also benefitted from several private communications from Milnor.

1. HYPERBOLIC SIMPLICES.

Throughout this chapter, k will denote an (Archimedean) ordered, square root closed field. We will use the projective model of hyperbolic space based on k. Other models are discussed in Thurston [68]; the presence of these different models leads to the interweaving of diverse areas of mathematics. For our purposes, the projective model appears to be the best.

For $n \in \mathbb{Z}^+$, we use $k^{1,n}$ to denote the k-vector space k^{1+n} of all column vectors of length 1+n so that it is equipped with the inner product $\langle\ ,\ \rangle_{1,n}$:

$$\langle u, v\rangle_{1,n} = {}^t u \cdot \mathrm{diag}(-1, I_n) \cdot v, \quad u, v \in k^{1+n}.$$

We continue to let $<\ ,\ >$ denote the standard positive definite inner product on k^{1+n}. The standard unit vectors will still be denoted by $e_0, \ldots, e_n$. The inner product $<\ ,\ >_{1,n}$ is therefore based on the quadratic form:

$$-X_0^2 + \Sigma_{1 \leq i \leq n} X_i^2.$$

In terms of the components $<v, e_i>$, we have:

$$<v, v>_{1,n} = -<v, e_0>^2 + \Sigma_{1 \leq i \leq n} <v, e_i>^2$$

There is no difficulty in passing to the direct limit by letting n go to ∞. We now consider the space of all rays $k^+ v$ in $k^{1,n}$, where $v \neq 0$.

The (finite) points of the hyperbolic n-space $\mathcal{H}^n = \mathcal{H}^n(k)$ are the rays $r = k^+ v$ characterized by the conditions:

$$< r, e_0 > = k^+ \text{ and } < r, r >_{1,n} = k^- = -k^+.$$

Each such ray $r = k^+ v$ contains a unique vector v with:

$$<v, e_0> > 0 \text{ and } < v, v >_{1,n} = -1.$$

When there is no chance of confusion, we also speak of the point v in $\mathcal{H}^n$. We note that $\mathcal{H}^n$ is an open convex subset of the space of all rays. When there is no chance of confusion, we will also view $\mathcal{H}^n$ as an open convex cone in $k^{1,n}$.

The infinite, or the boundary, points of $\mathcal{H}^n$ are the rays r in $k^{1,n}$ with:

$$<r, e_0> = k^+ \text{ and } <r, r>_{1,n} = 0.$$

These points form one component $\partial\mathcal{H}^n$ of the locus of the equation:

$$-X_0^2 + \Sigma_{1 \leq i \leq n} X_i^2 = 0.$$

The ultrainfinite points are the rays r in $k^{1,n}$ with:

$$< r, r >_{1,n} = k^+.$$

Each ultrainfinite point r contains a unique vector v in $k^{1,n}$ with:

$$r = k^+ v \text{ and } <v, v>_{1,n} = 1.$$

When there is no chance of confusion, we also speak of the ultrainfinite point v. The space of all ultrainfinite points is denoted by $U\mathcal{H}^n$. The space of all rays is therefore the disjoint union of $\mathcal{H}^n$, $\partial\mathcal{H}^n$, $U\mathcal{H}^n$, $-\partial\mathcal{H}^n$ and $-\mathcal{H}^n$. We note that the convex closure of $\partial\mathcal{H}^n$ is $\partial\mathcal{H}^n \cup \mathcal{H}^n$ while the convex closure of $U\mathcal{H}^n$ is the space of all rays. (If we view these as subsets of $k^{1,n}$, then the convex closure of $U\mathcal{H}^n$ has to include the empty ray $\phi = k^+ \cdot 0 = 0$.) $U\mathcal{H}^n$ is locally convex while $\mathcal{H}^n$ is such that the only nonempty convex subsets of $\mathcal{H}^n$ are reduced to points of $\mathcal{H}^n$.

The group of hyperbolic motions of $\mathcal{H}^n$ is denoted by $\Omega(1,n;k)$. It is the subgroup of index 2 in $O(<\, ,\, >_{1,n};k)$ preserving $\mathcal{H}^n$. We have:

$$O(<\, ,\, >_{1,n};k) = \Omega(1,n;k) \times \{\pm I_{n+1}\}.$$

The absence of $-I_{n+1}$ in $\Omega(1,n;k)$ is a major difference between the hyperbolic case and the spherical or the Euclidean case. In the latter two cases, -Id gave us a form of reduction from even dimensions to odd dimensions. The group $\Omega(1,n;k)$ has the subgroup $\Omega^+(1,n;k)$ of index 2 formed by all the transformations of determinant 1. This is then the subgroup of orientation preserving hyperbolic motions of $\mathcal{H}^n$. $\mathcal{H}^n$ can be identified with the coset space $\Omega(1,n;k)/O(n;k)$ or $\Omega^+(1,n;k)/SO(n;k)$. $\Omega(1,n;k)$ also acts transitively on $\partial\mathcal{H}^n$ and on $U\mathcal{H}^n$. As a variety, $\partial\mathcal{H}^n$ can be identified with spherical (n-1)-space. However, the geometry of $\partial\mathcal{H}^n$ is that of conformal geometry. If n = 2 or 3, $\partial\mathcal{H}^n$ can be identified with the projective line based on k and $k(\iota)$ respectively. The geometric algebra aspects of this assertion can be found in Artin [5]. $U\mathcal{H}^n$ can be identified with $\Omega(1,n;k)/\Omega(1,n-1;k)$ or with $\Omega^+(1,n;k)/\Omega^+(1,n-1;k)$. The action on $U\mathcal{H}^n$ can be extended to $O(<\, ,\, >_{1,n};k)$ so that we have two other identifications using O in place of Ω.

We can now raise the question of scissors congruence in each of the geometries $\mathcal{H}^n$, $\partial\mathcal{H}^n$, $U\mathcal{H}^n$. When n = 1, $U\mathcal{H}^1$ is actually isomorphic to $\mathcal{H}^1$; $\partial\mathcal{H}^1$ consists of two points. For $\mathcal{H}^1$, we can define 1-simplices by using convexity. We can also use the inner product $<\, ,\, >_{1,n}$ to define a k-valued invariant metric. It is then easy to check that length is a complete and independent invariant. The situation is completely similar to the spherical case. In essence, we are working with the function field $k(x,y)$, $x^2 - y^2 = 1$.

We next consider the cases for $n > 1$. The absence of convexity in $\partial\mathcal{H}^n$ prevents us from defining n-simplices through convexity. When $n = 2$ or 3, the group $\Omega^+(1,n;k)$ is respectively doubly and triply transitive on the points of $\partial\mathcal{H}^n$. As a consequence, for any reasonable scissors congruence data (e.g., an n-simplex is uniquely determined by its $n+1$ vertices) on the $(n-1)$-dimensional space $\partial\mathcal{H}^n$, $n = 2$ or 3, the group of stable scissors congruence classes must reduce to 0. A similar conclusion probably also holds for $n > 3$. The difficulty with $\partial\mathcal{H}^n$ can be attributed to the lack of an invariant volume for $n > 1$ (when k is Archimedean). We now consider $\mathcal{H}^n$ and $U\mathcal{H}^n$ separately.

As in the case of $n = 1$, $\mathcal{H}^n$ has an invariant k-valued metric. We can define hyperbolic n-simplices through convexity. To be precise, let $v_0, \ldots, v_n$ be k-linearly independent points of $\mathcal{H}^n$. Then:

$$\mathrm{hccl}\{v_0, \ldots, v_n\} = \{k^+ v \mid v = \Sigma_{0 \leq i \leq n} \alpha_i v_i \neq 0,\ \alpha_i \in k,\ \alpha_i \geq 0\}.$$

We note that each v above satisfies $< v, e_0 > > 0$ automatically and each k^+v singles out a unique vector v with $<v,v>_{1,n} = -1$. There is no problem in defining interior disjointness and subdivision and we have scissors congruence data. The group $\mathcal{P}(\mathcal{H}^n, \Omega(1,n;k))$ will be abbreviated to $\mathcal{P}\mathcal{H}^n$ or $\mathcal{P}\mathcal{H}^n(k)$.

$U\mathcal{H}^1$ can be identified with $\mathcal{H}^1$. For $n > 1$, we do not have either invariant metric or angle on $U\mathcal{H}^n$ in the usual sense. When k is Archimedean, we do have an invariant volume. As a result, we can hope for a nontrivial scissors congruence problem. We note that the important objects in a scissors congruence are the "small" n-simplices. Since we have local convexity and $O(<\ ,\ >_{1,n};k)$ is transitive on $U\mathcal{H}^n$, there is no problem in defining small n-simplices through local convexity. When k is Archimedean, Zylev's theorem holds for the scissors congruence data so defined. If we use radial projection from $0 \in k^{1,n}$ onto affine hyperplanes not passing 0, then line segments in $U\mathcal{H}^n$ and $\mathcal{H}^n$ actually appear as line segments. In general, each $u \in U\mathcal{H}^n$ is one of the two poles of a unique hyperplane in $\mathcal{H}^n$:

$$u^{\perp} \cap \mathcal{H}^n = \{k^+ v \in \mathcal{H}^n \mid <u,v>_{1,n} = 0\}.$$

We now describe the <u>small</u> <u>n-simplices</u> in $U\mathcal{H}^n$ precisely. Let $u_0, \ldots, u_n$

be k-linearly independent points in $U\mathcal{H}^n$. They are said to determine a small n-simplex in $U\mathcal{H}^n$ when the following condition holds:

$$\text{If } u = \Sigma_{0\leq i\leq n}\alpha_i u_i,\ \alpha_i \in k \text{ and } \Sigma_i \alpha_i = 1, \text{ then } < u, u >_{1,n} > 0 \qquad (1.1)$$

The small n-simplex $\mathrm{uccl}\{u_0, \ldots, u_n\}$ is then the collection of rays k^+u with u satisfying (1.1). For each such $\mathrm{uccl}\{u_0, \ldots, u_n\}$, we can associate the inner product matrix M in the space $M_{n+1}(k)$ of all $(n+1)\times(n+1)$ matrices:

$$M = (< u_i, u_j >_{1,n}),\ 0 \leq i, j \leq n. \qquad (1.2)$$

The k-linear independence of $u_0, \ldots, u_n$, the normalization $< u_i, u_i >_{1,n} = 1$ and condition (1.1) are then equivalent with:

(a) M is symmetric in $M_{n+1}(k)$ with diagonal I_{n+1};
(b) M has signature (1, n); and (1.3)
(c) ${}^t v \cdot M \cdot v > 0$ when v is nonzero in $k^{1,n}$ with $< v, e_i > \geq 0$, $0 \leq i \leq n$.

We note that $\mathrm{uccl}\{u_0, \ldots, u_n\}$ is a convex subset of $U\mathcal{H}^n$ and its vertices u_0, $\ldots, u_n$ are therefore uniquely determined up to an arbitrary permutation of order. The matrix M in (1.2) is therefore unique up to permutation equivalence (under conjugation by permutation matrices, or under simultaneous row and column permutations). Conversely, if $\mathrm{uccl}\{u_0, \ldots, u_n\}$ has the same inner product matrix as $\mathrm{uccl}\{v_0, \ldots, v_n\}$, then there is a unique σ in $O(< , >_{1,n}; k)$ so that $\sigma(u_i) = v_i$, $0 \leq i \leq n$; it follows that $\mathrm{uccl}\{u_0, \ldots, u_n\}$ is $O(< , >_{1,n}; k)$-congruent to $\mathrm{uccl}\{v_0, \ldots, v_n\}$. Finally, if $M \in M_{n+1}(k)$ satisfies (1.3), then the Sylvester law of inertia shows that we can find a nonsingular matrix A in $M_{n+1}(k)$ so that:

$${}^t A \cdot \mathrm{diag}(-1, I_n) \cdot A = M$$

If $u_0, \ldots, u_n$ denote the columns of A, then M is precisely the inner product matrix associated to $\mathrm{uccl}\{u_0, \ldots, u_n\}$. We summarize:

Proposition 1.4. Let k be an ordered, square root closed field. Then the inner product matrix correspondence establishes a bijection between the following objects:

(a) permutation equivalence classes of matrices satisfying (1.3);

(b) $O(<\,,\,>_{1,n};k)$-congruence classes of small n-simplices in $U\mathcal{H}^n$.

Remark 1.5. The preceding discussion can clearly be extended to points lying in $-\mathcal{H}^n \cup U\mathcal{H}^n \cup \mathcal{H}^n$ without any difficulty. If we want to include points on $-\partial\mathcal{H}^n \cup \mathcal{H}^n$, then the permutation equivalence must be replaced by monomial equivalence (congruent with respect to the normalizer in $GL(n+1;k)$ of the group of diagonal matrices having positive diagonal entries). We will use $gccl\{u_0, \ldots, u_n\}$ to denote the convex closure of the rays $k^+u_0, \ldots, k^+u_n$ in the space of all rays in $k^{1,n}$. Such an object will be called a generalized n-simplex. In reality, it is just an n-simplex in $S(k^{1+n})$. When all u_i lie in $U\mathcal{H}^n$, we also write $guccl\{u_0, \ldots, u_n\}$. When all u_i lie in $\mathcal{H}^n \cup \partial\mathcal{H}^n$, we also write $ghccl\{u_0, \ldots, u_n\}$; these are then called asymptotic or ideal n-simplices in $\mathcal{H}^n$ and the vertices lying on $\partial\mathcal{H}^n$ are also called cusps. The classification of the combinatorial types of generalized n-simplices is equivalent with the classification of the combinatorial types of placements of an Euclidean n-simplex in k^n with respect to the unit sphere $S(k^n)$. The following result is an evident modification of Proposition 1.4.

Proposition 1.6. Let k be an ordered, square root closed field. Then the inner product matrix correspondence establishes a bijection between:

(a) $\Omega(1,n;k)$- (resp. $O(n+1;k)$-) congruence classes of n-simplices in $\mathcal{H}^n(k)$ (resp. $S(k^{n+1})$);

(b) permutation equivalence classes of symmetric matrices M in $M_{n+1}(k)$ with diagonal entries -1 (resp. $+1$) such that all the principal minors of size $(i+1) \times (i+1)$ of M have signatures $(1,i)$ (resp. $(0,i+1)$), $1 \leq i \leq n$.

Remark 1.7. The process of simple subdivision can be described (in an unpleasant manner) in terms of the inner product correspondence. There are occasions where modified forms of (b) in Proposition 1.6 lead to interesting geometric objects obtained by performing some geometric operations on the associated generalized simplices.

In order to define Dehn invariants on $\mathcal{P}\mathcal{H}^n$, we need to describe the interior angles $\theta_A(A(I))$ at the face $A(I)$ of $A = \text{hccl}\{v_0, \ldots, v_n\}$, I any nonempty subset of $\{0, \ldots, n\}$. As in the spherical case, we define $u_0, \ldots, u_n$ in $k^{1,n}$ through the following formulas:

$$\begin{aligned} &< u_i, v_j >_{1,n} = 0 \text{ for } i \neq j \text{ and } < 0 \text{ for } i = j; \\ &< u_i, u_i >_{1,n} = 1 = - <v_i, v_j>_{1,n}; \text{ where } 0 \leq i, j \leq n \end{aligned} \tag{1.8}$$

Since the restriction of $< , >_{1,n}$ to $\text{Lsp}(A(\{i\}^c))$ has signature $(1, n-1)$, u_i exists and is one of the two possible poles of this hyperplane. The k-linear independence of $v_0, \ldots, v_n$ together with the nondegeneracy of $< , >_{1,n}$ then force $<u_i, v_i>_{1,n} \neq 0$. Condition (1.8) then picks out u_i uniquely. We call u_i the <u>exterior</u> <u>unit</u> <u>normal</u> of the codimensional 1 face $A(\{i\}^c)$ of A. The ordered set $u_0, \ldots, u_n$ is therefore uniquely determined by the ordered set $v_0, \ldots, v_n$. In general, $u_0, \ldots, u_n$ determines a generalized n-simplex $A^u = \text{guccl}\{u_0, \ldots, u_n\}$ in $U\mathcal{H}^n$. We call A^u the <u>hyperbolic</u> <u>dual</u> of A with the understanding that points of A^u may be outside of $U\mathcal{H}^n$. A^u can also be characterized as the convex cone:

$$A^u = \{k^+ u \mid <u, v>_{1,n} \leq 0 \text{ for all } k^+ v \text{ in } A \text{ and } u \neq 0\}.$$

We note that A^u can not contain both $k^+ w$ as well as $k^- w$ when $w \neq 0$ in $k^{1,n}$. The restriction of $< , >_{1,n}$ to $A^u(\{i\}^c)$ has signature $(0, n)$ for $0 \leq i \leq n$. We evidently have:

<u>Proposition 1.9</u>. Let k be an ordered, square root closed field. Then the inner product matrix correspondence establishes bijection between:

(a) $\Omega(1, n; k)$-congruence classes of hyperbolic duals of n-simplices in $\mathcal{H}^n$;

(b) permutation equivalence classes of symmetric matrices M in $M_{n+1}(k)$ with diagonal entries +1 so that M has signature $(1, n)$ but all $n \times n$ principal minors of M have signatures $(0, n)$.

If I is a nonempty subset of $\{0, \ldots, n\}$, then $I^c = \{0, \ldots, n\} - I$ is a proper subset of $\{0, \ldots, n\}$ and the face $A^u(I^c)$ can be viewed as a spherical simplex in $\text{Lsp}(A^u(I^c))$ with respect to the restriction of $< , >_{1,n}$. The interior angle $\theta_A(A(I))$ at the face $A(I)$ of A is defined to be the spherical dual $A^u(I^c)^{\S}$ in $S(\mathrm{Lsp}(A^u(I^c)))$. We note from (1.8) that $\mathrm{Lsp}(A^u(I^c))$ is the orthogonal complement of $\mathrm{Lsp}(A(I))$. Suppose that $I = \{0, \ldots, m\}$. Then $A^u(I^c)$ is just $\mathrm{sccl}\{u_{m+1}, \ldots, u_n\} = \mathrm{uccl}\{u_{m+1}, \ldots, u_n\}$ with $A^u(I^c)^{\S} = \mathrm{sccl}\{y_{m+1}, \ldots, y_n\}$:

$$\mathrm{Lsp}(A^u(I^c)) = \mathrm{Lsp}(A^u(I^c)^{\S}); \ < y_j, y_j >_{1,n} = 1;$$
$$< u_i, y_j >_{1,n} = 0 \text{ for } i \neq j \text{ and } < 0 \text{ for } i = j;\ m < i, j \leq n.$$

It is easy to see that $\mathrm{Lsp}(A(I)) + kv_i$ is the orthogonal sum of $\mathrm{Lsp}(A(I))$ and ky_i, $m < i \leq n$; in fact, y_i is on the same side as v_i with respect to the hyperplane $\mathrm{Lsp}(A(I))$ of $\mathrm{Lsp}(A(I)) + kv_i$. Geometrically, $A^u(I^c)^{\S}$ is just the collection of interior rays in the orthogonal complement of $\mathrm{Lsp}(A(I))$ bounded by the various codimensional 1 faces of A meeting at $A(I)$.

As done in the Euclidean case, the <u>total</u> <u>hyperbolic</u> <u>Dehn</u> <u>invariant</u> $H\Psi$ is given through the formula:

$$H\Psi[A] = \Sigma_I [A(I)] \otimes \overline{[\theta_A(A(I))]}, \text{ where } I \text{ ranges over all nonempty subsets of } \{0, \ldots, n\}, \tag{1.10}$$

As before, we form the direct sum $\mathcal{PH} = \coprod_{n \geq 0} \mathcal{PH}^n$ with the understanding that $\mathcal{PH}^0 = \mathbb{Z}\cdot[\text{point}] \cong \mathbb{Z}$ and that $[A]$ in (1.10) denotes the stable scissors congruence class of the n-simplex A in $\mathcal{PH}^n$. The argument in the spherical case can be imitated to show that we have an additive homomorphism:

$$H\Psi : \mathcal{PH} \longrightarrow \mathcal{PH} \otimes (\mathcal{PS}/\mathcal{CS}) \tag{1.11}$$

Moreover, $H\Psi$ is the structure map turning $\mathcal{PH}$ into a (right) comodule for the Hopf algebra $\mathcal{PS}/\mathcal{CS}$ and is compatible with the gradings so that we have subcomodules $\coprod_{0 \leq i \leq n} \mathcal{PH}^i$, $n \geq 0$. However, the signature $(1, n)$ prevents us from defining a product structure through the Minkowski orthogonal sum. On the other hand, we can try to extend the scissors congruence problem to $k^{p,q}$ for arbitrary signature (p, q). This will allow for a product structure. In order to make sure that we are not just dealing with the spherical case in disguise, the simplices have to be defined carefully. We will leave this generalization for another occasion. Instead, we will describe another plausible approach to the study of $\mathcal{PH}^n$.

The origin of this plausible approach is based on Gauss's proof of the defect formula for the area of a hyperbolic 2-simplex. A further push is supplied by the observation due to H. Kneser [46] that Schläfli's differential formula can be interpreted suitably for hyperbolic spaces. To be precise, a multiplicative factor of ι^n is needed, see Böhm [9] for more details. If n is even, this suggests that $\wp\mathcal{H}^n$ should be related to $\wp S^{n+1}$. In the spherical case, we saw that $\wp S^{2i+1} = [\text{point}] * \wp S^{2i}$ and this was related to the Gauss-Bonnet reduction from even dimension to odd dimension (recall that the degree in the spherical case is 1 + dimension). Finally, the Gauss-Bonnet theorem itself shows that the volume of a 2i-dimensional hyperbolic simplex can be expressed in terms of the various angles, see Allendorfer-Weil [1], Chern [20], Banchoff [7] among others. So far, all of these are accomplished on the level of volume. The idea is to try to achieve the result on the level of scissors congruence. In view of the fact that the general case has not been worked out in detail, we will limit ourselves to the case of $\mathcal{H}^2$ while the general case will be explored elsewhere, see Dehn [27].

Let $A = \text{hccl}\{v_0, v_1, v_2\}$ with $A^u = \text{guccl}\{u_0, u_1, u_2\}$. We now consider the following expression:

$$e[A] = \Sigma_i [A^u(\{i\}^c)] - [S(k^2)] \in \wp S^2 \tag{1.12}$$

The sum on the right hand side of (1.12) is the sum of the exterior angles at the vertices of A and is clearly dependent only on the $\Omega(1,2;k)$-congruence class of A. As it stands, this sum is not additive with respect to simple subdivision of A. For the case n = 2 under discussion, the additional term on the right hand side makes the expression additive with respect to simple subdivision of A. We therefore have an additive homomorphism:

$$e : \wp\mathcal{H}^2 \longrightarrow \wp S^2 \tag{1.13}$$

If we follow e by the volume map on $\wp S^2$, we get the defect formula:

> $\text{vol} \circ e[A] = 2 - (\alpha+\beta+\gamma)$, where $\alpha, \beta, \gamma \in k^+$ are the normalized interior angles at the vertices of A so that $[S(k^2)]$ corresponds to $4 = 2^2$.

We can now follow Gauss and prove $e \circ f = \mathrm{Id}$ on $\mathcal{P}S^2$. The proof is clear from the following beautiful picture due to Gauss:

We may now formulate our problem:

Combinatorial Gauss-Bonnet Problem. (Hyperbolic Case.) Find additive homomorphisms (if any):

$$e : \mathcal{P}(\mathcal{H}^{2i} \cup \partial\mathcal{H}^{2i}) \longrightarrow \mathcal{P}S^{2i}\text{; and}$$
$$f : \mathcal{P}S^{2i} \longrightarrow \mathcal{P}(\mathcal{H}^{2i} \cup \partial\mathcal{H}^{2i}),\ i \geq 0,$$

so that $e \circ f = \mathrm{Id}$ on $\mathcal{P}S^{2i}$ and so that the restriction of e to $\mathcal{P}\mathcal{H}^{2i}$ is compatible with all the Dehn invariants (perhaps even an isomorphism).

The use of asymptotic simplices in higher dimensional hyperbolic spaces also appeared in Böhm [9] and suggests the following problem:

Scissors Congruence Problem. (Idealized Hyperbolic Case.) Develope a theory for $\mathcal{P}(\mathcal{H}^n \cup \partial\mathcal{H}^n)$, $n \geq 0$.

Since all singly asymptotic 1-simplices are congruent, $\mathcal{P}(\mathcal{H}^1 \cup \partial\mathcal{H}^1) = 0$. This together with the absence of a Gauss-Bonnet theorem for odd dimensional manifolds (aside from the statement $0 = 0$) suggest that we probably should not expect to have a similar statement for the combinatorial Gauss-Bonnet problem with $2i+1$ in place of $2i$; similarly, $\mathcal{P}(\mathcal{H}^n \cup \partial\mathcal{H}^n)$ might turn out to be zero when n is odd. As noted, aside from the possible existence of torsion in $\mathcal{P}S^{2i}$, the spherical version of the combinatorial Gauss-Bonnet problem is solved for Archimedean k through the Hopf-Leray theorem. In particular, f is just the cone construction $*$[point]. However, we should probably not expect to have a similar reduction from $\mathcal{P}\mathcal{H}^{2i}$ to $\mathcal{P}\mathcal{H}^{2i-1}$. More precisely, $\mathcal{P}\mathcal{H}^2$ can be seen to be isomorphic to $\mathcal{P}S^2$ through e. However, $\mathcal{P}\mathcal{H}^1$ is isomorphic to k^+. To see this, we note that $\mathcal{H}^1$ can be identified with $\Omega^+(1, 1;k)$.

The elements of $\Omega^{+}(1,1;k)$ can be coordinatized as the collection of all pairs $\theta = (ch(\theta), sh(\theta))$, $ch(\theta) \in k^{+}$, $sh(\theta) \in k$, and $ch(\theta)^2 - sh(\theta)^2 = 1$. The product is given by the rule:

$$(a,b)\cdot(c,d) = (ac+bd, ad+bc),$$

Setting $u(\theta) = ch(\theta) + sh(\theta)$ and $v(\theta) = ch(\theta) - sh(\theta)$ shows that $\Omega(1,1;k) \cong k^{+}$. Since $\Omega^{+}(1,1;k)$ acts regularly on $\mathcal{H}^1$ and the simple subdivision process can be described directly as the product operation on k^{+}, we have $\mathcal{PH}^1 \cong k^{+}$. In analogy with the spherical case, an isomorphism between $\mathcal{PH}^1$ and $\mathcal{PH}^2$ ought not be just an abstract isomorphism. If it exists, it ought to be geometric. We already have the geometric isomorphism e between $\mathcal{PH}^2$ and $\mathcal{P}S^2$. This can be composed with the volume injection from $\mathcal{P}S^2$ to k when k is Archimedean. It follows that a geometric isomorphism between $\mathcal{PH}^1$ and $\mathcal{PH}^2$ would lead to an order preserving injective homomorphism:

$$\ell : k^{+} \longrightarrow k, \text{ k Archimedean.}$$

Up to a change of base, ℓ is just the logarithm map. A purely combinatorial map of this type should in principle be describable without using the completeness of the Archimedean field k. In particular, ℓ would make sense when k is the algebraic closure of $\mathbb{Q}$ in $\mathbb{R}$. This would imply that the value of the natural logarithm on positive real algebraic numbers ranges over a set of real numbers generating a field of transcendence degree at most 1 over $\mathbb{Q}$. This is a contradiction the theorem of Baker [6]. This is a loose argument indicative of the fact that we do not have a geometric construction similar to the cone construction in the spherical case. In fact, we have the following:

Construction Problem. Let k be an Archimedean ordered, square root closed field. Does there exist a family of polyhedra in $\mathcal{H}^n$, $n \geq 0$, such that:

(a) two polyhedra in the family are scissors congruent if and only if they have the same volume; and

(b) the set of volumes ahieved on this family is a dense subgroup of the image of the volume map on $\mathcal{PH}^n$.

Most of the problems posed in the spherical case have analogues in the hyperbolic case. We will refrain from carrying out the needed modification. Instead, we single out a few more scissors congruence problems.

Strong Scissors Congruence Problem. (Hyperbolic Case.) Let k denote an (Archimedean) ordered, square root closed field. Find enough reasonable (linear or semilinear) comodule homomorphisms to separate the points of the comodule $\mathcal{P}\mathcal{H}$ over the Hopf algebra $\mathcal{P}S/\mathcal{C}S$.

In a rough sense, the combinatorial Gauss-Bonnet problem is in this direction for the subcomodule $\mathcal{P}\mathcal{H}^{\mathrm{even}}$.

Conformal Scissors Congruence Problem. Does there exist scissors congruence data on $\partial\mathcal{H}^n$ with respect to the group of motions $\overset{+}{\Omega}(1,n;k)$ leading to a nontrivial group $\mathcal{P}(\partial\mathcal{H}^n, \overset{+}{\Omega}(1,n;k))$?

For the preceding problem, $\partial\mathcal{H}^n$ can be viewed as the (n-1)-dimensional sphere. The main problem is to find an appropriate definition of (n-1)-simplices bypassing the concept of convexity and then develope a nontrivial theory.

Ultrahyperbolic Scissors Congruence Problem. Develope a structure theory for the group $\mathcal{P}U\mathcal{H} = \coprod_{n\geq 0} \mathcal{P}U\mathcal{H}^n$ based on small simplices and on $O(1,\infty;k)$.

Indefinite Scissors Congruence Problem. Let $k^{\infty,\infty}$ be the direct limit of $k^{p,q}$ where $k^{p,q}$ is equipped with an inner product $\langle\ ,\ \rangle_{p,q}$ of signature (p,q). Define the simplices to be convex polyhedral cones in $k^{p,q}$ for varying p, q and let the group of motions be $O(\infty,\infty;k)$. Develope a structure theory for this scissors congruence data.

The preceding problem contains the spherical, the hyperbolic, the idealized hyperbolic, as well as the ultrahyperbolic cases. In this setting, we again have a product structure. However, the concepts of volume, angle and distance all become blurred. It is no longer clear what kind of a theory is reasonable. Perhaps the best that can be expected is some sort of interplay among the various subgroups or subrings.

Large Hyperbolic Simplex Problem. Let $k = \mathbb{R}$. An n-simplex in $\mathbb{H}^n \cup \partial\mathbb{H}^n$ is called regular if all its codimensional 2 dihedral angles are equal. Is it true that a regular totally asymptotic n-simplex in $\mathbb{H}^n \cup \partial\mathbb{H}^n$ has the largest volume among all the n-simplices of $\mathbb{H}^n \cup \partial\mathbb{H}^n$, $n > 1$?

The preceding problem was posed by M. Gromov. A discussion of its relevance can be found in Thurston [68; Chapter 6] in connection with some remarkable results of Gromov. Evidently, the critical n-simplices have to be totally asymptotic. The inner product matrix M associated to the dual A^u of a totally asymptotic n-simplex A can be characterized as follows:

(a) $M \in M_{n+1}(\mathbb{R})$ is symmetric with signature $(1, n)$;

(b) M has 1's on the diagonal; off diagonal entries are in $[-1, 1[$;

(c) the $(n-1)\times(n-1)$ principal minors of M are all positive definite; and

(d) the $n \times n$ principal minors of M all have determinant 0.

When $n = 2$, an affirmative answer follows from the fact that all triply asymptotic 2-simplices are congruent; it can also be seen from the Gauss-Bonnet formula or from the preceding characterization--all the off diagonal entries must be -1. When $n = 3$, an affirmative answer furnished by Milnor can be found in Thurston [68; Chapter 7]. In fact, Milnor posed two interesting conjectures about the values of the Lobachevsky function at algebraic points. Bloch and Wigner [8] have an elegant formula relating the volume of a totally asymptotic 3-simplex to dilogarithm of cross-ratios. In particular, there appears to be some interesting connection with algebraic K-theory, number theory and algebraic geometry. All these suggest the need for better understanding of the relevant analytic functions in volume calculations. Certain number theoretic aspects of these functions remind us of Hilbert's seventh problem (among others). We note also that the dilogarithm function appeared in combinatorial calculus of characteristic classes [29]. When $n = 4$, in a private communication, Milnor indicated a proof of a volume formula (attributed to Thurston) for a totally asymptotic hyperbolic 4-simplex. Presumably, an affirmative answer can be deduced from such a formula for $n = 4$.

2. RATIONAL SIMPLICES AND POLYHEDRA.

To simplify the discussion, we take $k = \mathbb{R}$. Consider a closed polyhedral ball in n-space (spherical, Euclidean, or hyperbolic). It is called rational if all the codimensional 2 dihedral angles are rational multiples of 2π in radian measure. In the spherical case, we indicated that these are possible sources of torsion in $\mathcal{P}S$. They can not be distinguished from elements of $\mathcal{P}S_n^{n+1}$ by the classical Dehn invariants. In all three cases, these belong in the kernel of the higher classical Dehn invariants. One source of these polyhedra originated with Schläfli when he calculated the volume of a number of rational spherical n-simplices. These turned out to be fundamental domains of finite Coxeter groups acting on n-spheres. The subset of these satisfying a crystallographic conditions is precisely the collection of Weyl groups in the classification of semisimple Lie algebras over $\mathbb{C}$. More generally, if we have a rational polyhedron such that each codimensional 2 dihedral angle is a submultiple of π in radian measure (i.e., has the form π/m for integer $m \geq 2$), then we can reflect the polyhedron with respect to its codimensional 1 faces. The product of two reflections with respect to two codimensional 1 faces meeting at a codimensional 2 face with interior dihedral angle π/m is then a rotation of order m. In this manner, we generate a Coxeter group with the given polyhedron as a fundamental domain (this assertion requires justification). As it stands, the Coxeter group is a discrete subgroup with compact quotient in the corresponding group of motions. In the spherical case, these are finite groups. In the Euclidean case, these are crystallographic groups. In general, these groups contain a subgroup of finite index acting freely on the associated spaces so that they lead to compact space forms, see Wolf [74]. The most interesting case appears in the hyperbolic spaces. In this case, we may allow the vertices of our polyhedron to lie on $\partial\mathcal{H}^n$. The associated Coxeter groups are then discrete subgroups in $O(1, n; \mathbb{R})$ with finite volume quotient. When $n \geq 2$, we have to allow for the possibility that $m = \infty$. We now provide a rough description in each of the cases. We will call these special polyhedra Coxeter polyhedra.

Spherical Case. The Coxeter polyhedra have been classified. They are orthogonal Minkowski sum of irreducible ones. The irreducible ones are simplices. Except when the rank (= 1 + dimension) is 2, there are only a finite number of irreducible ones in each rank. The classification is inductive. The point is that the interior angle at each vertex must again be a Coxeter simplex of one dimension lower. The procedure is carried out by means of the Coxeter diagram associated to the inner product matrix of the dual simplex. The ones that satisfy a crystallographic condition are precisely the ones that arose in Cartan's classification. The basic reference is Bourbaki [12].

Euclidean Case. The Coxeter polyhedra have been classified. They are orthogonal Minkowski sum of indecomposable ones (for the spherical case, indecomposable is equivalent with irreducible--the latter refers to a particular canonical representation of the Coxeter group). At each vertex, the interior angle corresponds to a spherical Coxeter simplex. The indecomposable ones (corresponding to simplices) are then classified by using the results from the spherical case.

Hyperbolic Case. Here we do not have the concept of orthogonal Minkowski sum. The canonical representation of the Coxeter group is always irreducible. We are not aware of a complete classification. However, a complete classification could presumably be achieved through the wealth of data contained in Vinberg [69], [70]. For example, the finite Coxeter simplices have been classified by Lanner [48], see also Bourbaki [12]. Similarly, the asymptotic Coxeter simplices have also been classified, see Koszul [47] as well as Bourbaki [12] and Vinberg [70]. These are related to the following problem: (see Mostow [54 ; 143-144].)

Discrete Subgroup Problem. Let Γ be a discrete subgroup of $\Omega(1,n;\mathbb{R})$ with finite volume quotient, is Γ necessarily arithmetic when n is large ?

$\Omega(1,n;\mathbb{R})$ can be replaced by other rank 1 Lie groups. Outside of examples for small values of n (arising from Coxeter polyhedra), there are no known general procedure to produce nonarithmetic Γ of the type mentioned.

In hyperbolic 3-space $\mathbb{H}^3$, Thurston [68; Chapter 7] has a procedure to construct Coxeter polyhedra. There is another method essentially described in the work of Vinberg. For the case of $\mathbb{H}^3$, it is also in the work of Andreev [2] and [3]. We will describe this procedure. We begin with an arbitrary finite hyperbolic 3-simplex. We then push the vertices towards $\partial\mathbb{H}^3$. When some or all the vertices reach $\partial\mathbb{H}^3$ (after travelling an infinite distance in the hyperbolic metric but only a finite distance in the ball model --the ball model is not an isometric model), we have asymptotic simplices. We continue pushing until the vertices lie in $U\mathbb{H}^3 \cup \mathbb{H}^3$ but not on $\partial\mathbb{H}^3$. For each vertex v in $U\mathbb{H}^3$, we form the polar hyperplane $\mathbb{R}v^\perp \cap \mathbb{H}^3$ to cut off the part of our generalized 3-simplex incident to v. When this is done carefully, we will have a polyhedron in $\mathbb{H}^3$ of the same combinatorial type as a Euclidean 3-simplex with some of its corners sliced off. At the truncated corners, the three dihedral angles are right angles. At the 6 original edges of our generalized 3-simplex, the interior dihedral angles have not changed. It is now apparent that the final polyhedron depends only on the hyperbolic dual of the original generalized 3-simplex. The truncation process makes sense for $\mathbb{H}^n$, $n \geq 2$. The difficulty lies in the construction of a generalized n-simplex with vertices in $U\mathbb{H}^n \cup \mathbb{H}^n$ so that all the codimensional 2 dihedral angles are submultiples of π in radians. We now consider some special cases corresponding to the case where all the vertices are in $U\mathbb{H}^n$, $n \geq 2$. Let the generalized n-simplex A have vertices $v_0, \ldots, v_n$ in $U\mathbb{H}^n$ so that all its faces of dimension $i > 0$ have nonempty intersection with $\mathbb{H}^n$. Let the dual simplex A^u have vertices $u_0, \ldots, u_n$ so that u_i must also lie in $U\mathbb{H}^n$. The truncating hyperplanes are $\mathbb{R}v_i^\perp = \Sigma_{j \neq i} \mathbb{R}u_j$ while the codimensional 1 faces lie in $\mathbb{R}u_i^\perp$. Let $M = (\langle u_i, u_j \rangle_{1,n})$, $0 \leq i, j \leq n$. The geometric conditions impose the following conditions on M:

(a) M is symmetric in $M_{n+1}(\mathbb{R})$ with 1's along the diagonal;

(b) M has signature $(1, n)$;

(c) all $n \times n$ principal minors have signatures $(1, n-1)$; all $(n-1) \times (n-1)$ principal minors are positive definite.

In order to obtain a Coxeter polyhedron at the end, we must impose the following integrality condition:

(d) if $n > 2$, then $\langle u_i, u_j \rangle_{1,n} = -\cos(\pi/m_{i,j})$, $m_{i,j}$ is an integer greater than 1 when $i \neq j$ ($m_{i,i} = 1$).

When $n = 2$, we obtain a 3 (continuous) parameter family of hyperbolic hexagons whose vertex angles are all equal to $\pi/2$ in radians. When $n = 3$, the conditions can be restated in terms of $(m_{i,j})$, $0 \leq i, j \leq 3$:

$$m_{i,j} = m_{j,i} \in \mathbb{Z}^+, \; m_{i,j} = 1 \text{ if and only if } i = j;$$

$$m_{p,q}^{-1} + m_{q,r}^{-1} + m_{r,p}^{-1} < 1, \text{ where } p, q, r \text{ are distinct.}$$

These lead to a 6 (discrete) parameter family of Coxeter polyhedra in $\mathbb{H}^3$ of the same combinatorial type as a Euclidean 3-simplex with all four corners sliced off. The 4 hexagonal faces have only right angles while the 4 triangular faces are all rational with angles given by $\pi/m_{p,q}$, $\pi/m_{q,r}$, and $\pi/m_{r,p}$ respectively. The main point is that the diophantine inequalities are not only neccessary but actually sufficient for the existence of a Coxeter polyhedron in $\mathbb{H}^3$ of the desired type.

When $n > 3$, the truncation process leads to a finite number of further Coxeter polyhedra. In general, Vinberg [69] has given precise criterion for the associated Coxeter groups to be arithmetic.

If we consider one of the inner product matrix M, then the associated Coxeter group is generated by the reflections with respect to the $2(n+1)$ hyperplanes $\mathbb{R}u_i^{\perp}$ and $\mathbb{R}v_j^{\perp}$, $0 \leq i, j \leq n$. They clearly leave invariant the quadratic form associated to the inner product matrix M. If we apply a Galois automorphism of $\mathbb{C}$ over $\mathbb{Q}$, the associated quadratic form may in some cases be positive definite. When this happens, the Coxeter group is conjugated to a subgroup of $O(4;\mathbb{R})$. The Coxeter group C(M) can be defined abstractly by generators and relations. The generators are just the $2(n+1)$ reflections ρ_i and σ_j mentioned before. The relations are of the form $\rho_i^2 = 1 = \sigma_j^2$ and $(\alpha\beta)^m = 1$ where π/m is the interior dihedral angle at a codimensional 2 face with α and β denoting the reflections with respect to two incident hyperplanes. From a theorem of Selberg [64], C(M) is known to have torsionfree subgroups of arbitrarily large index (C(M) is residually finite). In principle, it is possible to find such subgroups by mapping C(M) into finite groups along the line of Sah [58]. In the present case, it is easier to use reduction modulo high powers of prime ideals in the associated number fields. In any event, we can find torsionfree subgroups Γ so that Γ is normal and of finite index in C(M). When n = 3, our Γ then leads to a compact $K(\Gamma, 1)$ hyperbolic 3-manifold $\mathfrak{M}$ whose universal covering space is $\mathfrak{H}^3$. Taking a subgroup of index 2 if needed, we may take $\mathfrak{M}$ to be orientable. A fundamental domain for $\mathfrak{M}$ in $\mathfrak{H}^3$ is then made up from $|C(M):\Gamma|$ copies of the Coxeter polyhedron. It is now wellknown that the Eilenberg-MacLane homology and cohomology of Γ with local coefficients coincides with similar homology and cohomology of $\mathfrak{M}$. In particular, if $\mathfrak{M}$ is orientable, then $H_3(\Gamma, \mathbb{Z}) \cong \mathbb{Z}$. If we are in the case where C(M) can be conjugated by a Galois map σ into $O(4;\mathbb{R})$, then Γ can be assumed to be conjugated into $SO(4;\mathbb{R})$. This then sends $H_3(\Gamma, \mathbb{Z})$ into $H_3(SO(4, \mathbb{R}), \mathbb{Z})$ where $SO(4;\mathbb{R})$ is now viewed as a discrete group (!). We note that $\sigma(M)$ is the inner product matrix of a spherical 3-simplex dual to a raional 3-simplex (in general, not a Coxeter simplex). This is roughly the manner where the rational simplex problem arose in the work of Cheeger-Simons [19], the problem concerns the evaluation of a particular $\mathbb{R}/\mathbb{Z}$ valued cohomology class. In particular, it is unknown just what part of $H_3(SO(4, \mathbb{R}), \mathbb{Z})$ is described by $H_3(\Gamma, \mathbb{Z})$.

3. SCISSORS CONGRUENCE ON MANIFOLDS.

In private conversations, Milnor posed the question: What are some workable definitions for scissors congruences between compact Riemannian n-manifolds? In the case of space forms, the manifolds can be viewed as the quotient space $\overline{\mathfrak{M}}/\pi_1(\mathfrak{M}) = \mathfrak{M}$. Following Dirichlet's construction, we can find a polyhedral fundamental domain. As mentioned before, we can follow Siegel's argument to show that polyhedral fundamental domains are unique up to scissors congruence in $\overline{\mathfrak{M}}$. In this manner, the problem can be viewed as the determination of certain subgroup of the groups $\mathcal{P}^n(X,G)$ associated to each of the cases considered.

Theorem 3.1. Let $\mathfrak{m}$, $\mathfrak{n}$ be compact, flat, Riemannian n-manifolds so that their universal covering spaces are isometric to Euclidean n-space $\mathbb{R}^n$. Then $\mathfrak{m}$ and $\mathfrak{n}$ have scissors congruent fundamental polyhedra in $\mathbb{R}^n$ if and only if $vol_n(\mathfrak{m}) = vol_n(\mathfrak{n})$.

Proof. By the theorem of Bieberbach (connected with Hilberts eighteenth problem), $\mathfrak{m}$ and $\mathfrak{n}$ are finitely covered by flat tori. We can therefore find a positive integer N so that $N[\mathfrak{m}]$ and $N[\mathfrak{n}]$ lie in $\mathcal{P}E^n_n$. Since $\mathcal{P}E^n_n$ is a $\mathbb{Q}$-subspace of the $\mathbb{Q}$-vector space $\mathcal{P}E^n$ and volume defines an isomorphism from $\mathcal{P}E^n_n$ to $\mathbb{R}$, the assertion follows. Q.E.D.

Remark 3.2. The proof of Theorem 3.1 does not require the knowledge that $\mathcal{P}E^n$ is actually an $\mathbb{R}$-vector space in the present situation. In this respect, all the ingredients needed are contained in Hadwiger [31] plus the theorem of Bieberbach. Up to affine isomorphism, there are only a finite number of compact, flat, Riemannian manifold, see Charlap [16]. In dimensions n = 1, 2, 3 and 4, the classification is complete, the numbers are respectively 1, 2, 10, 74. In unpublished joint work with Charlap, a complete list (in the sense of presentations of the fundamental groups) has been compiled by hand. An "experimental" check shows agreement with the computer compilation made by H. Brown, J. Neubüser and H. Zassenhaus, see [14;p. 495]. We are indebted to Neubüser for the experimental checking carried out in private communications. More generally, by incorporating orientation, the Dehn invariants can be viewed as cochains. When viewed in this manner, they define cocycles. With the exception of volume, the higher Dehn invariants $\Phi^{(2i)}$, $i > 0$, define coboundaries--however, they are not the coboundaries of G-invariant cochains. In a manner similar to the analysis of higher differentials in the Hochschild-Serre spectral sequence, this also leads to cohomology of G (viewed as discrete groups). This is somewhat analogous to the syzygy problem. At any rate, the high Dehn invariants $\Phi^{(2i)}$, $i > 0$, must vanish on fundamental polyhedra of closed manifolds. These suggest the following conjecture:

<u>Scissors Congruence Conjecture for Manifolds</u>. (<u>Spherical and Hyperbolic Cases</u>.) Let $\mathfrak{M}$ and $\mathfrak{N}$ be compact n-dimensional Riemannian manifolds with common universal covering space isometric to the n-sphere or to the hyperbolic n-space. Then $\mathfrak{M}$ and $\mathfrak{N}$ have scissors congruent fundamental polyhedra in the universal covering space if and only if $\mathrm{vol}_n(\mathfrak{M}) = \mathrm{vol}_n(\mathfrak{N})$.

Since volume is a complete invariant for the corresponding scissors congruence problem when $n \leq 2$, we can restrict ourselves to $n > 2$ in the preceding conjecture. The necessity is clear. If we assume $\mathrm{vol}_n(\mathfrak{M})$ is given, then there are only a finite number of candidates for $\mathfrak{M}$. This is easy in the spherical case because $\mathrm{vol}_n(\mathfrak{M})$ determines the order of the fundamental group $\pi_1(\mathfrak{M})$ through the following formula:

$$|\pi_1(\mathfrak{M})| = \mathrm{vol}_n(S(\mathbb{R}^{n+1}))/\mathrm{vol}_n(\mathfrak{M}).$$

$\pi_1(\mathfrak{M})$ is necessarily a finite subgroup of $O(n+1;\mathbb{R})$ and its conjugacy class in $O(n+1;\mathbb{R})$ parametrizes $\mathfrak{M}$. It follows from the representation theory of finite groups that there are only a finite number of possibilities for $\mathfrak{M}$ when $\mathrm{vol}_n(\mathfrak{M})$ is specified. In the hyperbolic case, the finiteness result is much deeper. When $n > 3$, a theorem of Gromov asserts that there are only a finite number of possible hyperbolic n-manifolds of the type stated in the conjecture when the volume is bounded in advance. When $n = 3$, this strong finiteness result fails. However, a theorem of Jørgensen asserts that hyperbolic 3-manifolds of the type considered and with volume bounded in advance can be constructed out of a finite number of them through a sequence of hyperbolic Dehn surgery. It is our understanding that these Dehn surgeries may be viewed as scissors congruences on the associated fundamental polyhedra, for more details, see Thurston [68; Chapter 5]. We have now completed our rough trip by coming back to Dehn again.

Appendix

1. SHUFFLE ALGEBRA.

The basic references for this section are Milnor-Moore [52] and Sweedler [66]. Milnor-Moore worked in the graded category over a commutative ring. The Hopf algebras considered by them were not required to be either associative or coassociative. They also did not require the existence of an antipode. Sweedler worked in the category over a field, the existence of associative, coassociative laws as well as the existence of an antipode are part of the definition of a Hopf algebra. For our purposes, we follow the definition used by Sweedler except that the field is allowed to be a commutative ring. Ultimately, we will work over a field and assume a graded structure. Since the gradation will actually be even, graded commutativity is the same as commutativity. As a result, we can use the theorems from Milnor-Moore on graded commutative Hopf algebras.

Let V be a vector space (module) over a field (commutative ring) K. The shuffle algebra $Sh_K(V)$ is a commutative, graded, connected, Hopf algebra over K. The gradation is by nonnegative integers. Connectivity means that the degree 0 component is isomorphic to K through the unit as well as the counit (augmentation). As a K-vector space (module), $Sh_K(V)$ is isomorphic to the tensor algebra of V over K and inherits the grading by degree. The basic tensor product of $v_1, \ldots, v_n$ in V will be written as the finite sequence $(v_1, \ldots, v_n)$. This expresstion is therefore K-linear in each v_i. The empty sequence (.) plays the role of the identity element. The shuffle product is denoted by # so that $(u_1, \ldots, u_p)\#(v_1, \ldots, v_q)$ is the sum of all (p, q)-shuffles $(z_1, \ldots, z_{p+q})$ of the finite sequences $(u_1, \ldots, u_p)$ and $(v_1, \ldots, v_q)$. There are $(p+q)!/p!q!$ terms in this sum. The counit is simply the projection of $Sh_K(V)$ onto $Sh_K^0(V)$ with kernel $Sh_K^+(V) = \coprod_{i>0} Sh_K^i(V)$. The comultiplication (or diagonalization) Δ is the K-linear extension of the map sending the basic tensor product $(v_1, \ldots, v_n)$ onto the sum:

$$(.)\otimes(v_1, \ldots, v_n) + \ldots + (v_1, \ldots, v_j)\otimes(v_{j+1}, \ldots, v_n) + \ldots + (v_1, \ldots, v_n)\otimes(.)$$

Formally, the preceding sum can be viewed as the shuffle product of $\otimes$ and $(v_1, \ldots, v_n)$. With this interpretation, the coassociativity is immediate. In fact, the i-fold diagonalization amounts to the shuffle product of $(\otimes, \ldots, \otimes)$ (i factors) with $(v_1, \ldots, v_n)$. The antipode is the K-linear extension of the map sending the basic tensor product $(v_1, \ldots, v_n)$ onto $(-v_n, \ldots, -v_1)$ or $(-1)^n(v_n, \ldots, v_1)$. The remaining axioms of a Hopf algebra can be checked without difficulty. When K is a field, $Sh_K(V)$ is cocommutative if and only if $\dim_K V \leq 1$. When $\dim_K V$ is finite, $Sh_K(V)$ is the Hopf dual of the tensor algebra $T_K(V^*)$, $V^* = Hom_K(V, K)$. We recall that $T_K(V)$ is a Hopf algebra in which the comultiplication is the K-algebra homomorphism induced by sending each v in V onto $(.)\otimes v + v\otimes(.)$ in $T_K(V)\otimes T_K(V)$. The antipode is the K-linear extension of the map sending the basic tensor product $(v_1, \ldots, v_n)$ onto $(-1)^n(v_n, \ldots, v_1)$ again. When $\dim_K V$ is infinite, both $Sh_K(V)$ and $T_K(V)$ are the direct limit of respective Hopf subalgebras $Sh_K(W)$ and $T_K(W)$ with W ranging over the finite dimensional subspaces of V.

For any graded, connected, Hopf algebra C over the field K with Δ as the comultiplication, $P(C) = \{x \in C^+ \mid \Delta(x) = x\otimes 1 + 1\otimes x\}$ is called the space of primitive elements and $Q(C) = C^+/C^+ \cdot C^+$ is called the space of indecomposable elements. In general, these are graded K-vector spaces and P, Q are covariant functors of C. There is a natural map from P(C) to Q(C). The sequence of dimensions associated to Q(C) gives the minimum number of graded generators needed in each dimension in order to generate C as a graded K-algebra. The Hopf algebra C is said to be primitively generated when the natural map from P(C) to Q(C) is surjective. The primitive elements are said to be indecomposable when the natural map from P(C) to Q(C) is injective. The following basic facts are in Milnor-Moore [52]:

<u>Proposition 1.1</u>. Let C be a graded Hopf algebra over the field K so that $\dim_K C^i$ is finite for all i. Let C^* denote the Hopf dual of C so that C^{*i} is $Hom_K(C^i, K)$. Then $P(C^*)^i \cong Hom_K(Q(C)^i, K)$, $Q(C^*)^i \cong Hom_K(P(C)^i, K)$, $i \geq 0$.

Proposition 1.2. Let f: A ⟶ B be a homomorphism of graded, connected, Hopf algebras over a field K. Let P(f) and Q(f) be the induced maps. Then:

(a) f is injective if and only if P(f) is injective; and

(b) f is surjective if and only if Q(f) is surjective.

Proposition 1.3. Let C be a graded, connected, Hopf algebra over a field K of characteristic 0. Then:

(a) the primitive elements of C are indecomposable if and only if C is commutative; and

(b) C is primitively generated if and only if C is cocommutative.

Theorem 1.4. (Hopf-Leray) Let C be a graded, connected, Hopf algebra over a field K of characteristic 0. Assume C is graded commutative. As a K-algebra, C is isomorphic to the free, graded commutative K-algebra based on the graded K-vector space Q(C).

The two preceding results are proved in the graded category. If we artificially give our vector spaces an even grading, then graded commutativity is simply ordinary commutativity. For example, Theorem 1.4 implies that $Sh_K(V)$ is an integral domain when K is a field of characteristic 0.

Proposition 1.5. Let k be any ordered, square root closed field. The commutative ring $\mathbb{Q} \otimes (\mathcal{P}S/\mathcal{C}S)$ is isomorphic to the symmetric algebra over $\mathbb{Q}$ based on $\coprod_{i>0} \mathbb{Q} \otimes (\mathcal{P}S^{2i}/\mathcal{P}S_2^{2i})$. In particular, $(\mathcal{P}S/\mathcal{C}S)/\mathrm{Tor}(\mathcal{P}S/\mathcal{C}S)$ is an integral domain.

Proposition 1.6. Let k be an Archimedean ordered, square root closed field. Then $\mathcal{P}S/\mathrm{Tor}(\mathcal{P}S)$ is an integral domain.

The first assertion in Proposition 1.5 follows from Theorem 1.4 because $\mathbb{Q} \otimes (\mathcal{P}S/\mathcal{C}S)$ is a commutative, evenly graded, connected Hopf algebra over $\mathbb{Q}$. The second assertion follows from the first because $(\mathcal{P}S/\mathcal{C}S)/\mathrm{Tor}(\mathcal{P}S/\mathcal{C}S)$ is isomorphic to a subring of $\mathbb{Q} \otimes (\mathcal{P}S/\mathcal{C}S)$. For Proposition 1.6, we note from Proposition 3.22 of Chapter 6 that $S\Phi$ defines an injective ring homomorphism from $\mathcal{P}S/\mathrm{Tor}(\mathcal{P}S)$ into $\mathbb{R}[T] \otimes (\mathcal{P}S/\mathcal{C}S) \cong \mathbb{R}[T] \otimes_{\mathbb{R}} (\mathbb{R} \otimes (\mathcal{P}S/\mathcal{C}S))$. The latter is an integral domain by the same reasoning.

2. $\mathbb{Q}$-DIMENSION OF $\mathbb{Q}\otimes(\mathcal{P}S^{2i}/\mathcal{P}S_2^{2i})$, $i > 0$.

Throughout this section, k denotes an ordered, square root closed field. k is necessarily of characteristic 0 so that $|k|$ is infinite. It follows that:

$$\dim_{\mathbb{Q}} \mathbb{Q}\otimes(\mathcal{P}S^{2i}/\mathcal{P}S_2^{2i}) \leq |k|, \ i > 0.$$

For the reverse inequality, we use the ring homomorphism:

$$\mathbb{Q}\otimes\xi_2 : \mathbb{Q}\otimes\mathcal{P}S/\mathcal{C}S \longrightarrow \mathbb{Q}\otimes \mathrm{Sh}_{\mathbb{Z}}(\mathcal{P}S^2/\mathcal{P}S_2^2) \cong \mathrm{Sh}_{\mathbb{Q}}(\mathbb{Q}\otimes\mathcal{P}S^2/\mathcal{P}S_2^2).$$

Since $\mathbb{Q}\otimes\mathcal{P}S^2/\mathcal{P}S_2^2 \cong \mathbb{Q}\otimes\mathcal{P}S^2/4\mathcal{P}S_2^2$, we can use the identification of SO(2;k) $\cong U(1;k[\iota]) = \{a+\iota b \mid a, b \in k, \iota^2 = -1\}$ with $\mathcal{P}S^2/4\mathcal{P}S_2^2$. The reduced angle $\overline{\theta}$ is identified with $e(\theta) = c(\theta)+\iota s(\theta)$. In particular, if $\theta = [\mathrm{sccl}\{x_0, x_1\}]$, then $c(\theta) = \langle x_0, x_1\rangle$ and $s(\theta) = (1-c(\theta)^2)^{1/2} \in k^+$. The ambiguity drops out when we tensor with $\mathbb{Q}$. For Archimedean k, $e(\theta)$ is just $\exp(\iota\pi\theta/2)$.

A spherical (n-1)-simplex $\mathrm{sccl}\{x_0, \ldots, x_{n-1}\}$ is called <u>regular with parameter a</u> if $\langle x_i, x_j\rangle = a$ holds for all $i \neq j$. Such a simplex will be denoted by $SA_{n-1}\{a\}$. The k-linear independence of $x_0, \ldots, x_{n-1}$ and the positivity of $\langle\ ,\ \rangle$ translate to the positivity of the $n \times n$ symmetric matrix:

$$aJ_n + (1-a)I_n = (\langle x_i, x_j\rangle), \text{ where } I_n \text{ is the identity matrix and } J_n \text{ is the matrix with all entries equal to 1.}$$

The eigenvalues consist of $1+(n-1)a$ and $(1-a)$ with multiplicities 1 and n-1 respectively. The parameter a must necessarily satisfy:

$$-1/(n-1) < a < 1, \text{ where } n > 1.$$

Using the hypotheses on k together with the positivity of $\langle\ ,\ \rangle$, the preceding inequalities are sufficient to ensure the existence of $SA_{n-1}\{a\}$. Elementary calculation shows that the dual of $SA_{n-1}\{a\}$ is $SA_{n-1}\{-a/(1+(n-2)a)\}$ and:

$$\xi_2\overline{[SA_{2i-1}\{a\}]} = N_i(\overline{\theta}_1, \ldots, \overline{\theta}_i), \ i > 0, \text{ where } N_i \in \mathbb{Z}^+ \text{ and for } 1 \leq j \leq i,$$
$$c(\theta_j) = -a/(1+2(j-1)a), \ s(\theta_j) = [(1+2(j-1)a)^2 - a^2]^{1/2}/(1+2(j-1)a) \in k^+.$$

We will now show that for a suitable set of values of the parameter a, we have enough regular (2i-1)-simplices in $S(k^\infty)$ yielding $\mathbb{Q}$-linearly independent elements in $\mathbb{Q}\otimes(\mathcal{P}S^{2i}/\mathcal{P}S_2^{2i})$. First, we need a technical result:

Proposition 2.1. Let V be a vector space over a field K. Let $(v_{i,j})$ be an $m \times n$ matrix of elements from V satisfying the following conditions:

(a) for each i, $\dim_K \Sigma_{1 \leq j \leq n} Kv_{i,j} \geq \min(2, n)$; and

(b) for each (s, t), $v_{s,t} \notin \Sigma_{1 \leq i \leq s-1} \Sigma_{1 \leq j \leq n} Kv_{i,j}$.

Then the images of $x_i = (v_{i,1}, \ldots, v_{i,n})$ in $Q(Sh_K(V))^n$ are K-linearly independent, $1 \leq i \leq m$.

Proof. When n = 1, $Q(Sh_K(V))^1 = V$ and the assertion follows from (b). We may therefore assume $n \geq 2$ so that min(2, n) = 2. Since the assertion is of finite character, we may assume $\dim_K(V) < \infty$. The hypotheses are clearly inductive, we may therefore assume that $x_1, \ldots, x_{m-1}$ have K-linearly independent images in $Q(Sh_K(V))^n$. By way of contradiction, assume:

$$x_m = \Sigma_{1 \leq i \leq m-1} \alpha_i x_i + z, \ \alpha_i \in K \text{ and } z \text{ is decomposable.}$$

Since $\dim_K(V) < \infty$, $T_K(V^*)$ is the Hopf dual of $Sh_K(V)$. The space $P(T_K(V^*))$ of primitive elements in $T_K(V^*)$ is a Lie subalgebra of $T_K(V^*)$ and the dual of $Q(Sh_K(V))^n$ can be identified with $P(T_K(V^*))^n$ for each n. If $f_1, \ldots, f_n$ lie in V^*, then $(f_1, \ldots, f_n) \in T_K^n(V^*)$ is identified with an element of the dual of $Sh_K^n(V)$ through the formula:

$$(f_1, \ldots, f_n)\{(v_1, \ldots, v_n)\} = f_1(v_1) \cdot \ldots \cdot f_n(v_n), \ v_i \in V.$$

Let $W = \Sigma_{1 \leq i \leq m-1} \Sigma_{1 \leq j \leq n} Kv_{i,j}$. We can identify $(V/W)^*$ with a K-subspace of V^*. We therefore have:

$$0 = [f, g](x_m) = f(v_{m,1})g(v_{m,2}, \ldots, v_{m,n}) - f(v_{m,n})g(v_{m,1}, \ldots, v_{m,n-1})$$

where $f \in (V/W)^* \subset V^* = P(T_K(V^*))^1$ and $g \in P(T_K(V^*))^{n-1}$.

If $v_{m,1}$ and $v_{m,n}$ are K-linearly independent mod W, then the preceding equation together with the duality between $P(T_K(V^*))^{n-1}$ and $Q(Sh_K(V))^{n-1}$ imply that $(v_{m,2}, \ldots, v_{m,n}) = (v_{m,1}, \ldots, v_{m-1,n}) = 0$. Since $n > 1$, this contradicts (b). If $v_{m,1}$ and $v_{m,n}$ are K-linearly dependent mod W, then (b) implies that $v_{m,1}+W$ and $v_{m,n}+W$ are distinct from W and are nonzero multiples of each other. Multiplying by suitable elements of $K^{\times}$, we may assume that $v_{m,1}+W = v_{m,n}+W$. The exhibited equation together with the duality between $P(T_K(V^*))^{n-1}$ and $Q(Sh_K(V))^{n-1}$ now imply:

$$(v_{m,2}, \ldots, v_{m,n}) = (v_{m,1}, \ldots, v_{m,n-1}) \text{ in } Sh_K(V)$$

From hypothesis (b), we have: $Kv_{m,1} = Kv_{m,2} = \ldots = Kv_{m,n} \neq 0$. Since $n > 1$, this contradicts (a).

Q.E.D.

<u>Proposition 2.2</u>. Let k be an ordered, square root closed field. Then:

$$\dim_{\mathbb{Q}} \mathbb{Q} \otimes (\wp S^{2i}/\wp S_2^{2i}) = |k|, \ i > 0$$

<u>Proof</u>. We first consider the case $i = 1$. We need to find $e(\theta_\lambda) = a_\lambda + \iota b_\lambda$ with a_λ, $b_\lambda \in k$ so that $a_\lambda^2 + b_\lambda^2 = 1$ and so that $e(\theta_\lambda)$ are multiplicatively independent in $k[\iota]$ for λ ranging over an index set of cardinality equal to $|k|$. There are two subcases.

Subcase 1. k is uncountable. The transcendence degree of k over Q must be equal to $|k|$. Multiplying by -1 and taking reciprocal when needed, we can find a transcendence base a_λ for k over Q so that $0 < a_\lambda < 1$. The corresponding $e(\theta_\lambda)$ with $b_\lambda = (1-a_\lambda^2)^{1/2} \in k^+$ then form a transcendence base for $k[\iota]$ over $\mathbb{Q}$ so that they are multiplicatively independent.

Subcase 2. k is countable. Let $a_\lambda = (u_\lambda^2 - v_\lambda^2)/(u_\lambda^2 + v_\lambda^2)$, $b_\lambda = 2u_\lambda v_\lambda/(u_\lambda^2 + v_\lambda^2)$ with $u_\lambda > v_\lambda > 0$ represent relatively prime integers. It is well known that every prime p congruent to 1 mod 4 has the form $X^2 + Y^2$ for suitable X, Y integer in $\mathbb{Z}$ and that there exists an infinite number of such primes. It is also well known that $\mathbb{Z}[\iota]$ is a unique factorization domain. If we let u_λ, v_λ be chosen to range over those pairs with $u_\lambda^2 + v_\lambda^2$ representing all primes congruent to 1 mod 4, then $e(\theta_\lambda) = \bar{z}_\lambda / z_\lambda$ where $\bar{z}_\lambda \cdot z_\lambda$ is the unique factorization of the rational prime $u_\lambda^2 + v_\lambda^2$ into distinct primes in $\mathbb{Z}[\iota]$. The multiplicative independence of $e(\theta_\lambda)$ is then a consequence of the unique factorization theorem in $\mathbb{Z}[\iota]$.

We next consider the case $i > 1$. In $\mathbb{Q} \otimes (\wp S/\mathcal{C}S)$, the space of indecomposables is isomorphic to $\coprod_{i>0} \mathbb{Q} \otimes (\wp S^{2i}/\wp S_2^{2i})$. Since the ring homomorphism carries this space into $Q(Sh_K(V))$, we can use Proposition 2.1 to check $\mathbb{Q}$-linear independence of the classes $\overline{[SA_{2i-1}\{a\}]}$ in $\mathbb{Q}\otimes(\mathcal{P}S^{2i}/\mathcal{P}S_2^{2i})$. According to our earlier calculation, we need to examine elements of the form:

$$(\overline{\theta}_1, \ldots, \overline{\theta}_i) \in Sh^i_{\mathbb{Q}}(\mathbb{Q}\otimes \mathcal{P}S^2/\mathcal{P}S_2^2), \text{ where } c(\theta_j) = -a/(1+2(j-1)a),$$
$$s(\theta_j) = [(1+2(j-1)a)^2 - a^2]^{1/2}/(1+2(j-1)a), \ 1 \le j \le i, \text{ and } -1/(2i-1) < a < 1.$$

If we let $a = y/x$, $0 < y < x$ in k, then $c(\theta_j) = -y/(x+2(j-1)y)$, and $s(\theta_j) = [x+(2j-1)y]^{1/2}[x+(2j-3)y]^{1/2}/(x+2(j-1)y)$. We again have two subcases:

Subcase 1. k is uncountable. As in the case $i = 1$, we let $0 < y_\lambda < x_\lambda$ so that $\{x_\lambda, y_\lambda \mid \lambda \in \Lambda\}$ form a transcendence base of k over $\mathbb{Q}$. Since k is not countable, $|k| = |\Lambda|$. We assert that $(\overline{\theta_1(\lambda)}, \ldots, \overline{\theta_i(\lambda)})$ have $\mathbb{Q}$-linearly independent images in $Q(Sh_{\mathbb{Q}}(\mathbb{Q}\otimes(\mathcal{P}S^2/\mathcal{P}S_2^2)))^i$, where $\theta_j(\lambda)$ corresponds to substituting (x_λ, y_λ) for (x, y) in representing a as y/x. The assertion is of finite character, we may therefore restrict λ to a finite subset of Λ. The condition (b) of Proposition 2.1 is a simple consequence of the algebraic independence of $\{x_\lambda, y_\lambda \mid \lambda \in \Lambda\}$. We note that the $\mathbb{Q}$-linearly independence among the $\overline{\theta}$'s translates to the multiplicative independence of the corresponding $e(\theta)$'s. Since $i > 1$, condition (a) of Proposition 2.1 is satisfied if we can show the multiplicative independence of $e(\theta_1(\lambda))$ and $e(\theta_2(\lambda))$. Each of these determine a quadratic extension field of $\mathbb{Q}(x_\lambda, y_\lambda)$. Since each has infinite order, multiplicative dependence forces the equality of the extension. This means that $(x-y)(x-3y)$ is a perfect square in $\mathbb{Q}(x, y)$ where x, y are algebraically independent over $\mathbb{Q}$. Since $\mathbb{Q}[x, y]$ is a unique factorization domain and $(x-y)$, $(x-3y)$ are nonassociate irreducible elements, we must have multiplicative independence of $e(\theta_1(\lambda))$ and $e(\theta_2(\lambda))$. As a result:

> The classes $\overline{[SA_{2i-1}\{y_\lambda/x_\lambda\}]}$ are $\mathbb{Q}$-linearly independent in $\mathbb{Q}\otimes(\mathcal{P}S^{2i}/\mathcal{P}S_2^{2i})$, where $0 < y_\lambda < x_\lambda$ so that $\{x_\lambda, y_\lambda \mid \lambda \in \Lambda\}$ forms a transcendence base for k over $\mathbb{Q}$.

Subcase 2. k is countable. We proceed as in the preceding case. This time $0 < y_\lambda < x_\lambda$ will be selected to be suitable elements of $\mathbb{Z}$. Each $e(\theta_j(\lambda))$

determines an imaginary quadratic extension field of $\mathbb{Q}$. We will select x_λ and y_λ inductively to satisfy the following conditions:

(1) x_λ, $y_\lambda \in \mathbb{Z}^+$ so that $x_\lambda > y_\lambda$, $\lambda \in \mathbb{Z}^+$;

(2) for each $m \in \mathbb{Z}^+$, y_m is twice the product of all rational prime divisors of the intergers $(x_\lambda + 2(j-1)y_\lambda)$, $1 \leq \lambda \leq m-1$;

(3) x_λ is an odd prime greater than $2(i-1)y_\lambda$.

If $e(\theta_j(\lambda))$ were a root of 1, then $c(\theta_j(\lambda))$ must be half of an integer. Since y_λ is even and $(x_\lambda + 2(j-1)y_\lambda)$ is odd, this is never the case. Thus, $e(\theta_j(\lambda))$ is not a root of 1. Since $e(\theta_j(\lambda))$ has norm 1, it is not an algebraic integer and the unique factorization of the principal ideal generated by $e(\theta_j(\lambda))$ in its quadratic field can involve only the prime ideal divisors of $x_\lambda + 2(j-1)y_\lambda$. By the choice made in (2) and (3), every prime ideal divisor of $x_\lambda + 2(j-1)y_\lambda$, $1 \leq \lambda \leq m-1$, $1 \leq j \leq i$, must be distinct from every prime ideal divisor of $x_m + 2(t-1)y_m$, $1 \leq t \leq i$, in any number field containing all these expressions. In terms of $(\overline{\theta_1(\lambda)}, \ldots, \overline{\theta_i(\lambda)})$, this translates into condition (b) of Proposition 2.1. It is also immediate that $x_\lambda - y_\lambda$ and $x_\lambda - 3y_\lambda$ are relatively prime so that we can proceed as in the preceding subcase to conclude that condition (a) of Proposition 2.1 also holds. As a result:

The classes $[\overline{SA_{2i-1}\{y_\lambda/x_\lambda\}}]$ are $\mathbb{Q}$-linearly independent in $\mathbb{Q} \otimes (\mathcal{P}S^{2i}/\mathcal{P}S_2^{2i})$, where x_λ, y_λ satisfy conditions (1), (2) and (3) listed above.

Combining these cases, we have shown $\dim_{\mathbb{Q}} \mathbb{Q} \otimes (\mathcal{P}S^{2i}/\mathcal{P}S_2^{2i}) \geq |k|$ for $i > 0$. As mentioned before, the reverse inequality is elementary. Q.E.D.

Proposition 2.3. Let k be an ordered, square root closed field. Then:

$\dim_{\mathbb{Q}} Q(\mathbb{Q} \otimes \mathcal{P}S)^i$ is 1, 0 or $|k|$ according to i is 1, odd and $\neq 1$, or even and positive.

Proof. $Q(\mathbb{Q} \otimes \mathcal{P}S)^i$ is respectively $\mathbb{Q}[\text{point}]$, 0 or $\mathbb{Q} \otimes (\mathcal{P}S^i/\mathcal{P}S_2^i)$. Q.E.D.

We can apply these results to the Euclidean case. The argument is similar.

We use the k-algebra homomorphism:

$$(\text{gr. vol.} \otimes \xi_2) \circ E\Psi : \wp E \longrightarrow k[T] \otimes Sh_{\mathbb{Z}}(\wp S^2/\wp S^2_2)$$

The space $Q(\wp E) = \wp E^+/\wp E^+ \cdot \wp E^+$ is naturally isomorphic to $\coprod_{i \geq 0} \mathcal{G}E_1^{2i+1}$. On $\mathcal{G}E_1^{2i+1}$, $(\text{gr. vol.} \otimes \xi_2) \circ E\Psi$ amounts to the evaluation of the map $(\text{gr. vol.} \otimes \xi_2) \circ E\Psi^{(2i)}$ on all of $\wp E^{2i+1}$. It follows that:

$$\dim_k \mathcal{G}E_1^{2i+1} \geq \dim_k (\text{gr. vol.} \otimes \xi_2) \circ E\Psi^{(2i)}(\wp E^{2i+1}), \ i \geq 0.$$

When $i = 0$, $\wp E^1 = \mathcal{G}E_1^1 \cong k$ through the graded volume map. We therefore consider the case $i > 0$. For each $\lambda \in k^+$, let $EB_n\{\lambda\}$ denote the n-dimensional Euclidean crossed polytope $\text{ccl}\{\pm e_1, \ldots, \pm e_{n-1}, \pm \lambda e_n\}$, where $e_1, \ldots, e_n$ denote an orthonormal set in k^∞. Straightforward calculation shows:

$$(\text{gr. vol.} \otimes \xi_2) \circ E\Psi^{(2i)}[EB_{2i+1}\{\lambda\}]$$
$$= T \otimes \{c_n(\overline{\theta}_3, \ldots, \overline{\theta}_{2i-1}, \overline{\alpha(2i+1, \lambda)}) + (z_{i-1}(\lambda), \overline{\beta(2i+1, \lambda)})\}$$

where $c_n \in k^+$, $z_{i-1}(\lambda) \in Sh_k^{i-1}(k \otimes \wp S^2/\wp S^2_2)$;

$c(\theta_j) = 1/j$, $j = 3, 5, \ldots, 2i-1$;

$e(2\alpha(2i+1, \lambda)) = \{1 + \lambda(-2i)^{1/2}\}^2/\{2i\lambda^2+1\}$; and

$e(2\beta(2i+1, \lambda)) = \{\lambda + (\lambda^2(1-2i)-1)^{1/2}\}^2/\{2i\lambda^2+1\}$.

Since $2c(\theta_j)$ is not an algebraic integer, $e(\theta_j)$ is not a root of 1. Thus, $\overline{\theta}_3, \ldots, \overline{\theta}_{2i-1}$ are nonzero when viewed as elements in $k \otimes \wp S^2/\wp S^2_2$. A k-linear dependence relation among the classes of $[EB_{2i+1}\{\lambda\}]$ mod $\wp E_2^{2i+1}$ (equivalently, among the components of $[EB_{2i+1}\{\lambda\}]$ in $\mathcal{G}E_1^{2i+1}$), implies a linear dependence relation among the elements $\overline{\alpha(2i+1, \lambda)}$ and $\overline{\beta(2i+1, \lambda)}$ viewed as elements of $\mathbb{Q} \otimes \wp S^2/\wp S^2_2$. This translates into a multiplicative dependence relation among the elements $e(2\alpha(2i+1, \lambda))$, $e(2\beta(2i+1, \lambda))$. As in the proof of Proposition 2.2, the argument can be divided into two cases:

Case 1. k is uncountable. Let λ range over a transcendence base of k over $\mathbb{Q}$. A nontrivial multiplicative dependence relation among a finite set of $e(2\alpha(2i+1, \lambda))$, $e(2\beta(2i+1, \lambda))$ must imply a nontrivial multiplicative dependence relation between $e(2\alpha(2i+1, \lambda))$ and $e(2\beta(2i+1, \lambda))$. It is immediate that $c(\alpha(2i+1, \lambda))$ and $c(\beta(2i+1, \lambda))$ are both transcendental over $\mathbb{Q}$ when

$i > 0$. It follows that neither $e(2\alpha(2i+1,\lambda))$ nor $e(2\beta(2i+1,\lambda))$ is of finite order when $i > 0$. Each of these leads to an imaginary quadratic extension field of $\mathbb{Q}(\lambda)$. A nontrivial multiplicative dependence relation between $e(2\alpha(2i+1,\lambda))$ and $e(2\beta(2i+1,\lambda))$ would force the equality of discriminants. This is equivalent with $(-2i)(\lambda^2(1-2i)-1)$ being a perfect square in $\mathbb{Q}(\lambda)$. This is evidently not the case when $i > 0$.

Case 2. k is countable. Since k is square root closed and $i > 0$, we can use Dirichlet's theorem on arithmetic progressions to find an infinite set of $\lambda \in k^+$ so that $\lambda^2 \in \mathbb{Z}^+$ and $2i\lambda^2+1$ ranges over an infinite set Λ of primes in $\mathbb{Z}^+$. By extracting square roots, the corresponding $2c(\alpha(2i+1),\lambda)$ and $2c(\beta(2i+1),\lambda)$ are both algebraic but not integral over $\mathbb{Z}$. It follows again that neither $e(2\alpha(2i+1),\lambda)$ nor $e(2\beta(2i+1,\lambda))$ is of finite order when $2i\lambda^2+1$ is a prime in Λ. A nontrivial multiplicative dependence relation among a finite set of $e(2\alpha(2i+1),\lambda)$, $e(2\beta(2i+1),\lambda)$ leads to a nontrivial multiplicative dependence relation among the corresponding principal (fractional) ideals in an algebraic number field of finite degree over $\mathbb{Q}$. Since $e(2\alpha(2i+1),\lambda)$ and $e(2\beta(2i+1),\lambda)$ both have norm 1 with algebraic integral numerator and having the same denominator $2i\lambda^2+1$, the corresponding principal ideals must involve only prime ideal divisors of the rational prime $2i\lambda^2+1$. The unique factorization theorem for ideals shows that a nontrivial multiplicative dependence relation must imply a nontrivial dependence relation involving the principal fractional ideals associated to $e(2\alpha(2i+1,\lambda))$ and $e(2\beta(2i+1),\lambda)$ for a single λ so that $2i\lambda^2+1$ is a prime in Λ. If they determine distinct imaginary quadratic extension, then we can find a Galois automorphism fixing $e(2\alpha(2i+1),\lambda)$ and inverting $e(2\beta(2i+1),\lambda)$. Combining this with the nontrivial dependence relation, we would conclude that $e(2\alpha(2i+1,\lambda))$ and $e(2\beta(2i+1,\lambda)$ are both units. This contradicts the fact that they are not integral over $\mathbb{Z}$. Consequently, they must determine the same imaginary quadratic extension field. Up to a square factor in $\mathbb{Z}^+$, the discriminant is $-2i\lambda^2$. It follows that the rational prime $2i\lambda^2+1$ becomes the product of two conjugate prime ideals say $\mathfrak{p}$ and $\bar{\mathfrak{p}}$ in the common imaginary quadratic extension field of $\mathbb{Q}$. Since $e(2\alpha(2i+1),\lambda)$ and $e(2\beta(2i+1),\lambda)$ both have norm 1, their numerators therefore generate either $\mathfrak{p}^2$ or $\bar{\mathfrak{p}}^2$. If $2i\lambda^2+1 > 3$, ± 1 are the only units in our imaginary quadratic extension of $\mathbb{Q}$. We can therefore conclude that $s(2\alpha(2i+1,\lambda))^2 = s(2\beta(2i+1,\lambda))^2$, or $-2i\lambda^2 = \lambda^2(1-2i)-1$. This forces $\lambda^2 = 1$ or $\lambda = 1$.

Putting all these together, we have shown:

If $i > 0$, then the classes $[EB_{2i+1}\{\lambda\}]$ are k-linearly independent in $\mathcal{P}E^{2i+1}/\mathcal{P}E_2^{2i+1} \cong \mathcal{G}E_1^{2i+1}$ when either of the following conditions hold:

(1) $\lambda \in k^+$ ranges over a transcendence base of k over $\mathbb{Q}$;

(2) $\lambda \in k^+$ ranges over the elements with $\lambda^2 \in \mathbb{Z}$, $\lambda > 1$ and $2i\lambda^2+1$ a prime in $\mathbb{Z}^+$.

As a consequence, we have:

Proposition 2.4. Let k be an ordered, square root closed field. Then:

$\dim_k \mathcal{G}E_1^{2i+1}$ is 1 or $|k|$ according to $i = 0$ or > 0.

The preceding result can be extended easily to yield:

Proposition 2.5. Let k be an ordered, square root closed field. Then:

$\dim_k \mathcal{G}E_i^n$ is 0, 1 or $|k|$ according to $i \not\equiv n \bmod 2$, $i = n$, or $i \equiv n \bmod 2$ and $1 \leq i < n$.

Remark 2.6. Since the graded volume map is 0 on $\mathcal{G}E_i^n$ for $i \neq n$, Propositions 2.4 and 2.5 yield affirmative answers to the original third problem of Hilbert in a strong sense. In a similar way, Proposition 2.3 is the spherical analogue. These results indicate that $\mathcal{P}S$ and $\mathcal{P}E$ are huge as rings and k-algebras respectively. In the analysis of the crossed polytopes $EB_{2i+1}\{\lambda\}$ for the case where k is countable, we could have restricted to λ in $\mathbb{Q}^+$. This would have involved an extended version of Dirichlet's theorem on arithmetic progressions--specifically, it involves the representations of primes by primitive, indecomposable, binary, integral quadratic forms, see Weber [72; vol. 3, book 2, 404-405]. The present argument avoids this more delicate result. According to the theorem of Lindemann, $\pi\alpha(2i+1,\lambda)$ and $\pi\beta(2i+1,\lambda)$ are real transcendental numbers for the algebraic λ's considered by us. According to the theorem of Baker [6 ; Theorem 2.1], the linear independence of these numbers over $\mathbb{Q}$ actually imply the linear independence of these numbers over the algebraic closure $\overline{\mathbb{Q}}$ of $\mathbb{Q}$. It would follow from the yet to be decided conjecture of Schanuel that these numbers are actually algebraic independent over $\mathbb{Q}$. These interesting questions are connected with the seventh problem of Hilbert, see Baker [6] and the survey article of Tijdeman [14; 241-268]. We are indebted to James Ax for a number of interesting conversations on these topics. In particular, some of the dimensional calculations were simplified by astute observations made by him.

Similar calculations can be made in the hyperbolic case. In the absence of a multiplicative structure, we can only make assertions about the size of $\mathbb{Q} \otimes \mathcal{P}\mathcal{H}^n$ in terms of its $\mathbb{Q}$-dimension. We can work with regular simplices just as in the spherical case. The only change needed is the restriction on the value of the parameter a to: $a < -1/n$ for the case of a regular hyperbolic n-simplex. Since there is no multiplicative structure to worry about, the calculations are much simpler. We get $\dim_{\mathbb{Q}} \mathbb{Q} \otimes \mathcal{P}\mathcal{H}^n$ equals to 1 or cardinality of k according to $n = 0$ or > 0. We omit the details.

References

1. C. B. Allendorfer and A. Weil, The Gauss-Bonnet theorem for Riemannian polyhedra, Trans. Amer. Math. Soc., 53 (1943), 165-179.

2. E. M. Andreev, On convex polyhedra in Lobacevskii spaces, Math. USSR-Sbornik, 10 (1970), 413-440.

3. E. M. Andreev, On convex polyhedra of finite volume in Lobacevskii space, Math. USSR-Sbornik, 12 (1970), 255-259.

4. K. Aomoto, Analytic structure of Schläfli function, Tokyo Univ., 1976, preprint.

5. E. Artin, Geometric Algebra, Interscience, New York, 1957.

6. A. Baker, Transcendental Number Theory, Cambridge Univ. Press, 1975.

7. T. Banchoff, Critical points and curvature for embedded polyhedra, J. Diff. Geom., 1 (1967), 245-256.

8. S. Bloch, Higher regulators, algebraic K-theory, and zeta functions of elliptic curves, Univ. of Chicago, 1978, preprint.

9. J. Böhm, Untersuchung des Simplexinhaltes in Räumen konstanter Krümmung beliebiger Dimension, J. Reine Angew. Math., 202 (1959), 16-51.

10. J. Böhm, Zu Coxeters Integrationsmethode in gekrümmten Räumen, Math. Nachr., 27 (1963), 179-214.

11. V. G. Boltianskii, Hilbert's Third Problem, Wiley, New York, 1978.

12. N. Bourbaki, Groupes et Algebres de Lie, Hermann, Paris, 1968, Chapters 4, 5 and 6.

13. R. Bricard, Sur une question de geometrie relative aux polyedres, Nouv. Ann. Math., 15 (1896), 331-334.

14. F. E. Browder, Mathematical Developments Arising from Hilbert Problems, Proc. Symp. Pure Math., 28 (1976), Amer. Math. Soc., Providence.

15. H. Cartan and S. Eilenberg, Homological Algebra, Princeton Univ. Press, 1956.

16. L. S. Charlap, Compact, flat Riemannian manifolds, I, Ann. of Math., 81 (1965), 15-30.

17. J. Cheeger, A combinatorial formula for Stiefel-Whitney classes, Topology of Manifolds, Markham, Chicago, 1970, 470-471.

18. J. Cheeger, Invariants of flat bundles, Proc. Int. Cong. of Math., Vancouver, (1974), 3-6.

19. J. Cheeger and J. Simons, Differential characters and geometric invariants, SUNY Stony Brook, 1973, preprint.

20. S. S. Chern, A simple intrinsic proof of the Gauss-Bonnet formula for closed Riemannian manifolds, Ann. of Math., 45 (1944), 747-752.

21. S. S. Chern, On the curvatura integra in a Riemannian manifold, Ann. of Math., 46 (1945), 674-684.

22. H. S. M. Coxeter, The functions of Schläfli and Lobatschefsky, Quart. J. of Math., 6 (1935), 13-29.

23. H. S. M. Coxeter, Non-Euclidean Geometry, Univ. of Toronto Press, 1968, 3rd ed.

24. H. S. M. Coxeter, Regular Polytopes, Dover, New York, 1973, 3rd ed.

25. M. Dehn, Über den Rauminhalt, Math. Ann., 55 (1901), 465-478.

26. M. Dehn, Über den Inhalt spharischer Dreiecke, Math. Ann., 60 (1905), 166-174.

27. M. Dehn, Die Eulersche Formel im Zusammenhang mit dem Inhalt in der Nicht-Euklidischen Geometrie, Math. Ann., 61 (1906), 561-586.

28. H. E. Debrunner, Zerlegungsrelationen zwischen regulären Polyedern des E^d, Arch. d. Math., 30 (1978), 656-659.

29. A. M. Gabrielov, I. M. Gel'fand and M. V. Losik, Combinatorial computation of characteristic classes, Func. Anal. and Appl., 9 (1975), 103-115, 186-202.

30. C. F. Gauss, Werke, 8 (Grundlägen der Geometrie), Teubner, Leipzig, 1900, 159-270.

31. H. Hadwiger, Vorlesungen über Inhalt, Oberfläche und Isoperimetrie, Springer-Verlag, Berlin, 1957.

32. H. Hadwiger, Neuere Ergebnisse innerhalb der Zerlegungstheorie euclidscher Polyeder, Jber. Deutsch. Math.-Verein. 70 (1967/68), 167-176.

33. H. Hadwiger, Translative Zerlegungsgleichheit der Polyeder des gewohnlicher Raumes, J. Reine Angew. Math., 233 (1968), 200-212.

34. H. Hadwiger and P. Glur, Zerlegungsgleichheit ebener Polygone, Elem. Math., 6 (1951), 97-106.

35. S. Halpern and D. Toledo, Stiefel-Whitney classes, Ann. of Math., 96 (1972), 511-525.

36. B. Harris, Group cohomology classes with differential form coefficients, Algebraic K-theory, Evanston 1976, Lect. Notes in Math., 551, Springer-Verlag, Berlin-Heidelberg-New York, 1976, 278-282.

37. D. Hilbert, Grundlägen der Geometrie, Teubner, Leipzig, 1968, 10 ed.

38. G. Hochschild, Introduction to Affine Algebraic Groups, Holden Day, San Francisco, 1971.

39. G. Hochschild and J.-P. Serre, Cohomology of group extensions, Trans. Amer. Math. Soc., 74 (1953), 110-134.

40. B. Jessen, The algebra of polyhedra and the Dehn-Sydler theorem, Math. Scand., 22 (1968), 241-256.

41. B. Jessen, Zur Algebra der Polytope, Göttingen Nachr. Math. Phys. (1972), 47-53.

42. B. Jessen, Einige Bemerkungen zur Algebra der Polyeder in nicht-Euclideschen Räumen, Comm. Math. Helv., 53 (1978), 525-528.

43. B. Jessen, J. Karpf and A. Thorup, Some functional equations in groups and rings, Math. Scand., 22 (1968), 257-265.

44. B. Jessen and A. Thorup, The algebra of polytopes in affine spaces, (to appear).

45. I. J. Kaplansky, Hilbert's Problems, Lect. Notes in Math., Univ. of Chicago, 1977, (preliminary ed.).

46. H. Kneser, Der Simplexinhalt in der nichteuklidischen Geometrie, Deut. Math., 1 (1936), 337-340.

47. J. L. Koszul, Lectures on hyperbolic Coxeter groups, Univ. of Notre Dame, 1967/68.

48. F. Lanner, On complexes with transitive groups of automorphisms, Comm. Sem. Math. Univ. Lund. 11 (1950).

49. L. Lewin, Dilogarithms and Associated Functions, MacDonald, London, 1958.

50. S. MacLane, Homology, Springer-Verlag, Berlin, 1963.

51. W. Magnus, Max Dehn, Math. Intelligencer, 1 (1978), 132-143.

52. J. W. Milnor and J. C. Moore, On the structures of Hopf algebras, Ann. of Math., 81 (1965), 211-264.

53. E. E. Moise, Elementary Geometry from an Advanced Viewpoint, Addison Wesley, Reading, 1963.

54. G. D. Mostow, Discrete subgroups of Lie groups, Queen's Papers in Pure and Appl. Math., 48 (1978), 65-153.

55. D. Mumford, Introduction to Algebraic Geometry, Lect. Notes, Harvard University, c. 1965.

56. O. Nicolletti, Sulla equivalenze dei poliedri, Rend. Circ. Mat. Palermo, 40 (1915), 194-210.

57. C. H. Sah, Automorphisms of finite groups, J. Algebra, 10 (1968), 47-68, 44 (1977), 573-575.

58. C. H. Sah, Groups related to compact Riemann surfaces, Acta Math., 123 (1969), 13-42.

59. C. H. Sah, Cohomology of split group extensions, J. Algebra, 29 (1974), 255-302, 45 (1977), 17-68.

60. C. H. Sah, Hilbert's Third Problem, Lecture Notes, SUNY Stony Brook, 1977.

61. C. H. Sah and J. B. Wagoner, Second homology of Lie groups made discrete, Comm. in Algebra, 5 (1977), 611-642.

62. L. A. Santalo, Integral Geometry and Geometric Probability, Ency. Math. and its Appl. 1, Addison Wesley, Reading, 1976.

63. L. Schläfli, Gesam. Math. Abhand., 1, Birkhauser, Basel, 1950.

64. A. Selberg, Harmonic analysis and discontinuous groups in weakly symmetric Riemannian spaces with application to Dirichlet series, J. Ind. Math. Soc., 20 (1956), 47-87.

65. C. L. Siegel, Discontinuous groups, Ann. of Math., 44 (1943), 674-689.

66. M. Sweedler, Hopf Algebras, Benjamin, New York, 1969.

67. J. P. Sydler, Conditions necessaires et suffisantes pour l'equivalence des polydres l'espace euclidien a trois dimensions, Comm. Math. Helv., 40 (1965), 43-80.

68. W. Thurston, The geometry and topology of 3-manifolds, Lect. Notes, Princeton Univ., 1977/78.

69. E. B. Vinberg, Discrete groups generated by reflections in Lobacevskii spaces, Math. USSR-Sbornik, 1 (1967), 429-444.

70. E. B. Vinberg, Some arithmetical discrete groups in Lobacevskii spaces, Discrete Subgroups of Lie Groups and Applications to Moduli, Oxford Univ. Press, 1975, 324-348.

71. W. Wallace, Leybourne's Mathematical Repository, 3, 1807.

72. H. Weber, Lehrbuch der Algebra, 3, Chelsea, New York, 1908, 3rd. ed.

73. H. Whitney, Tensor products of abelian groups, Duke Math. J., 4 (1938), 495-528.

74. J. A. Wolf, Spaces of Constant Curvature, Publish or Perish, Boston, 1974, 3rd ed.

75. M. Zacharias, Elementar Geometrie und Elementar Nicht-Euklidsche Geometrie in Synthetischer Behandlung, Ency. d. Math. Wiss., III, 1-2, 859-1172.

76. V. B. Zylev, Equicomposability of equicomplementable polyhedra, Sov. Math. Doklady, 161 (1965), 453-455.

77. V. B. Zylev, G-composedness and G-complementability, Sov. Math. Doklady, 179 (1968), 403-404.

Items 13, 56 and 71 were not directly accessible to us. We learned of their approximate contents through secondary sources. In addition, the following references as well as others (not yet accessible to us) may already contain some if not all the ideas mentioned in Chapter 8 pertaining to $\mathcal{PH}^n$.

H. Liebmann, Nichteuklidischen Geometrie, Leipzig, 1912.

N. J. Lobatschefskij, Zwei geometrische Abh., Kasaner Bote, 1829/30.

N. J. Lobatschefskij, Imaginare Geometrie, Kasaner gelehrte Schr. 1836.

Index